6TH EDITION

DEVELOPING SKILLS IN ALGEBRA
A LECTURE WORKTEXT

J. Louis Nanney
Miami-Dade Community College

John L. Cable
Miami-Dade Community College

Linda L. Tully
University of Pittsburgh at Johnstown

Theresa M. Shustrick
University of Pittsburgh at Johnstown

Joe Wilson
University of Pittsburgh at Johnstown

HAWKES
LEARNING
SYSTEMS

Editor: Harding Brumby
Development Director: Marcel Prevuznak
Layout: QSI (Pvt.) Ltd.: U. Nagesh, E. Jeevan Kumar, D. Kanthi, B. Syam Prasad, K. V. S. Anil
Art: Ayvin Samonte

HAWKES
LEARNING
SYSTEMS

A division of Quant Systems, Inc.

Library of Congress Control Number: 2006921438

Printed in the United States of America

ISBN:
978-1-932628-22-7
1-932628-22-3

CONTENTS

CHAPTER 6   Exponents and Radicals   245

CHAPTER 7   Quadratic Equations   297

CHAPTER 8   Relations and the Cartesian   347
Coordinate System

PREFACE

J. Louis Nanney and John L. Cable created *Developing Skills in Algebra: A Lecture Worktext* for the student who needed a comprehensive introduction to or review of basic algebraic skills. The text was written in a style that was meant to be easily accessible to the student. We adopted their worktext for use in our Algebra 1 course when the course was developed, and we continued to use every edition of it since then.

In the years since we first decided to use *Developing Skills in Algebra*, the demographics of our student population changed, and the courses for which Algebra 1 is a prerequisite (college algebra, liberal arts math courses, general education math courses, and math for elementary teachers) were restructured. We used approximately twelve different titles by various authors for these subsequent courses. We continue to use Nanney and Cable's text for our introductory level course because it invariably prepares our students for their next course.

In 2005, we asked Hawkes Learning Systems if a more current edition would soon be available. The skills Nanney and Cable presented so clearly did not change over the years, but the applications were no longer current. We learned that both Nanney and Cable are deceased and that Hawkes would be willing to work with us on a new edition.

Why Did We Choose to Update This Text?

We believe that as instructors of mathematics, we have three responsibilities to our students: 1) help them build a strong foundation of skills and a clear understanding of concepts in mathematics, 2) bring them to an appreciation of the beauty of mathematics, and 3) encourage them to continue in the study of mathematics. Since we teach mostly freshman-level courses, our efforts are concentrated on the first responsibility because students must first learn the concepts and skills before they can see the patterns within the concepts and the relationships among them.

Developing Skills in Algebra provides a solid foundation that serves students well in subsequent courses which require knowledge of basic algebra. Students needing a refresher course, as well as those who have not previously been exposed to algebra, benefit from its use. It works when used in a traditional setting and can be used as a self-paced tutorial or in a math lab setting. Most college algebra texts begin with a review chapter of the skills and concepts contained in Chapters 1 through 6 of this text. For a more traditional mathematics curriculum, we find these chapters provide an excellent preparation for a college algebra course.

Students often find reading any math text a daunting process, yet our students often tell us *this is one text they can understand.* Each section contains clear and concise explanations of definitions, rules, and procedures followed by detailed examples. An intuitive approach rather than one of rigorous proof is used. The pages are uncluttered, free of extraneous information, and can be read top to bottom.

What Features Did We Keep?

We maintained the style of the text. Each chapter starts with a **Chapter Survey** that can be used to help students and instructors identify concepts on which the student should concentrate. Instruction can be tailored to the strengths and weaknesses identified by the Survey results. All answers are provided in the answer key, and each is coded to the appropriate section.

Each section begins with the **Objectives** for that section. The objectives indicate what skill(s) the student will be learning.

Sections of the chapters are frequently divided into **subsections** followed by exercises designed to allow immediate practice of skills. In this way, the students are building a solid foundation of basic skills.

Definitions, rules, and **procedures** are enclosed in boxes.

Each concept and skill is rigorously developed using **numerous examples** that build on each other and increase gradually in level of difficulty.

Common student errors are identified and set apart as **Cautions**. They are highlighted in red to help students find and correct errors.

The Section **Exercises** provide practice on every skill and concept presented in the section. The answers to the odd problems are provided in the **answer key**.

Each chapter is followed by a **Summary** that lists the terminology, rules, formulas, and procedures covered in the chapter, with sections indicated for easy reference.

A **Chapter Review** follows the Summary. This review consists of problems for all the skills and concepts in the chapter. Answers are referenced to the section number.

A **Practice Test** is provided for each chapter to help students prepare for examinations. Each test covers key concepts in the chapter. Again, the answers are coded to the section in which the concept appears.

A **Mid-Book Test** and an **End-of-Book Test** are included.

What Did We Change?

We *expanded explanations* of some concepts and *included new explanations* of others where necessary so that rules, definitions, and procedures are justified. Students can better remember what to do if they have some understanding of why they are doing it.

We *changed the format of the examples* to clearly state the directions. By doing so, the students will learn what the directions for the section exercises are asking. In the solution of an example problem, we *provided an explanation for each step* prior to or to the left of the actual algebraic manipulation. This way, the students will continue to read top to bottom and left to right as they read through the solution. Although we still develop each concept using numerous examples, we *adapted the progression of examples*. The first examples for the concept now show and explain all steps. We then gradually omit explanations until we give an example that reads as we would expect a completed student solution to be written.

We included *additional Cautions* to point out common student errors.

We included *additional definition, rule, and procedure boxes* for concepts already in the text but not set apart.

We added *color as a teaching tool* to highlight important steps in solutions so that students will clearly see what rules are used.

We added *more diagrams and illustrations* to aid visual learners in their understanding of formulas and in their translation of problems into algebraic equations.

We added more *problems that contain or result in rational numbers* throughout the text, beginning with the first section, to help students build number sense and confidence in working with numbers that are not integers. To this end, we encourage our students to do many of the problems without the use of a calculator.

We changed the discussion of word problems in Chapter 3 to include *Polya's approach to problem solving*. Most of the problems in this chapter were changed for better readability and ease of translation to algebra. Students should find these problems more accessible.

At the end of sections that present several concepts in subsections, we included a *Practice Your Skills* set of exercises. These problems are designed to use all of the concepts discussed in the section and sometimes concepts from previous sections, thus illustrating the cumulative nature of algebra.

We included more examples and problems in Chapter 6 to help students *build skills with rational exponents and radicals*. These are two of the topics in the review chapter of college algebra texts with which students struggle.

If you are a past user of *Developing Skills in Algebra*, you will also notice some rearrangement of topics in Chapters 2, 4, 7, and 8.

Linda Tully
Terry Shustrick
Joe Wilson

Acknowledgements

We extend a sincere thank you to our spouses (John, Jim, and Theresa) and children (especially Sara) for their patience and understanding. The user of this text will meet them as he/she progresses through the text since we have used their names in some of our problems.

We also want to thank, for their insights and suggestions, the other instructors in our department who use this text. In particular, we thank Beth Hoffman and Dawn Cable for reviewing a draft of each chapter. In addition, we thank our department work-study students for working problems and checking answers.

We are indebted to all those who in some way contributed to the first five editions of this text.

Last but definitely not least, we offer a very special thank you to Hawkes Learning Systems, especially Harding Brumby (editor), for working with us to produce this edition.

CHAPTER 1

SURVEY

The following questions refer to material discussed in this chapter. Work as many problems as you can and check your answers with the answer key in the back of the book. The results will direct you to the sections of the chapter in which you need to work. If you answer all questions correctly, you may already have a good understanding of the material contained in this chapter.

1. Classify $\sqrt{5}$ as rational or irrational.

1. _____

2. Give the negative or opposite of (-31).

2. _____

3. Evaluate: $|-4|$

3. _____

4. Evaluate: $(-94) + 47$

4. _____

5. Evaluate: $-20 - (-14)$

5. _____

6. Evaluate: $8 + (-35) - 4$

6. _____

7. Evaluate: $(-14)(-3)$

7. _____

8. Evaluate: $\dfrac{-36}{6}$

8. _____

9. Evaluate: $(-3)(5)(-2)(-1)$

9. _____

10. Simplify: x^4x^3

10. _____

11. Multiply: $\left(7a^2b\right)\left(-5a^3b^4\right)$

11. _____

12. Divide: $\dfrac{27x^8y}{-3x^2y^3}$

12. _____

13. Combine similar terms:
$5x^2y - 2xy + 7xy^2 - 4x^2y - 3xy$

13. _____

14. Multiply: $-3a\left(a^2 - 2a + 1\right)$

14. _____

15. Evaluate: $7 + 4 \cdot 2 - 5$

15. _____

16. Simplify: $-x^2 + 4x\left[2x - \left(x - y\right)\right]$

16. _____

17. Evaluate: $3x^2 - 4x + 2$ when $x = -2$

17. _____

Fundamental Concepts

Certain fundamental ideas form the foundation of the study of algebra and of algebraic manipulations. It is important that the student of algebra be proficient in the skills of arithmetic.

The topics to be covered in this chapter include the operations on signed numbers and operations on monomials and polynomials, including the technique of evaluating expressions. Mastering these skills is necessary to the further study of algebra.

1–1   ADDITION OF REAL NUMBERS

OBJECTIVES

Upon completion of this section you should be able to:
1. Give the negative or opposite of a number.
2. Find the absolute value of a number.
3. Add real numbers.

The numbers first encountered in elementary arithmetic are those used in counting or the set of **counting numbers**, $\{1, 2, 3, 4, \ldots\}$. This set is later extended to include zero and is then called the **whole numbers**. The whole numbers along with the negatives of the counting numbers give us the set of **integers**,

$$\{\ldots, -4, -3, -2, -1, 0, 1, 2, 3, 4, \ldots\}.$$

The set of **rational numbers** is the set of all numbers that can be expressed as a ratio of two integers. These numbers are generally referred to as **fractions**. This set includes the integers. As an example, 3 can be expressed as $\frac{3}{1}, \frac{6}{2}$, and so on. In fact, this set includes all numerical expressions that involve only the operations of addition, subtraction, multiplication, and division.

The set of **irrational numbers** is a set of numbers that cannot be expressed as the ratio of two integers. This set includes such numbers as $\pi, \sqrt{5}$, and $\sqrt[3]{7}$.

The sets of rational and irrational numbers together make up a set called the **real numbers**. Elementary algebra is a study of the real numbers and their properties.

EXERCISE 1-1-1

Classify each of the following numbers as counting, whole, integer, rational, or irrational. A number may have more than one classification.

1. -2

2. 0

3. $\dfrac{2}{3}$

4. $\dfrac{-10}{7}$

5. $\dfrac{8}{4}$

6. $\dfrac{-25}{5}$

7. $\sqrt{3}$

8. π

The basic set of properties of the real numbers can be used to justify all of the manipulations used in both arithmetic and elementary algebra. If this set of properties is accepted as being true then all other properties can be proved as theorems. Although an intuitive approach rather than one of rigorous proof is used in this text, mention is often made of these basic properties.

Properties of Real Numbers

1. *The real numbers are closed under addition.* If a and b are real numbers, then $(a + b)$ is also a real number. Since 2 and 3 are real numbers, $(2 + 3)$ is also a real number.

2. *Addition of real numbers is commutative.* If a and b are real numbers, then $(a + b) = (b + a)$. Since order does not matter, $(2 + 3) = (3 + 2)$.

3. *Addition of real numbers is associative.* If a, b, and c are real numbers, then $(a + b) + c = a + (b + c)$. Since grouping does not matter, $(2 + 3) + 4 = 2 + (3 + 4)$.

4. *There is a real number that is an additive identity. This number is zero.* For all real numbers a, $a + 0 = a$. Zero added to any real number gives the original number. So $2 + 0 = 2$.

5. *Every real number has an additive inverse.* If a is a real number, then there is a real number $(-a)$ such that $a + (-a) = 0$. Adding a real number to its opposite will always result in zero. So $2 + (-2) = 0$.

6. *The real numbers are closed under multiplication.* If a and b are real numbers, then $(a \cdot b)$ is also a real number. Since 2 and 3 are real numbers, $(2 \cdot 3)$ is also a real number.

7. *Multiplication of real numbers is commutative.* If a and b are real numbers, then $a \cdot b = b \cdot a$. Since order does not matter, $2 \cdot 3 = 3 \cdot 2$.

8. *Multiplication of real numbers is associative.* If a, b, and c are real numbers, then $(a \cdot b)c = a(b \cdot c)$. Since grouping does not matter, $(2 \cdot 3)4 = 2(3 \cdot 4)$.

9. *There is a real number that is a multiplicative identity. This number is 1.* For all real numbers a, $a \cdot 1 = a$. Multiplying a real number by 1 will always result in the original number. So $2 \cdot 1 = 2$.

10. *Each nonzero real number has a reciprocal (multiplicative inverse).* If a is a nonzero real number, then there is a real number $\frac{1}{a}$ such that $a \cdot \frac{1}{a} = 1$. Multiplying a nonzero real number by its reciprocal will always result in 1. So $2 \cdot \frac{1}{2} = 1$.

11. *The real numbers obey the distributive property of multiplication over addition.* If a, b, and c are real numbers, then $a(b + c) = ab + ac$. So $2(3 + 4) = (2 \cdot 3) + (2 \cdot 4)$.

These properties are listed on the inside front cover for easy reference.

Members of the set of real numbers can also be classified as positive, negative, or zero. Numbers such as $+7, +\frac{3}{4}$, and $+10$ are positive, while $-3, -\frac{4}{5}$, and -5 are negative. Positive numbers can be designated by a $(+)$ sign and negative numbers are designated by a $(-)$ sign. Such numbers are often referred to as **signed numbers**. If a nonzero number is not preceded by a sign, it is understood that the number is positive. So numbers such as $7, \frac{3}{4}$, and 10 are positive numbers.

All numbers greater than zero are **positive** and all numbers less than zero are **negative**. Zero is neither positive nor negative.

It is important to distinguish between a negative number and the negative (also called the opposite or additive inverse) of a number. The **negative (opposite** or **additive inverse)** of a given number is a number that when added to the given number yields zero. This is the *additive inverse property* for real numbers, listed as Property 5 for real numbers on the front cover.

Example 1   The opposite of $+3$ is -3 because $+3 + (-3) = 0$. ∎

Example 2   The opposite of -3 is $+3$ because $-3 + (+3) = 0$. ∎

EXERCISE 1-1-2

Give the negative (opposite or additive inverse) of each.

1. $+5$ 2. 18 3. -2 4. $\frac{3}{4}$

5. $-1\dfrac{1}{2}$ **6.** $-\dfrac{5}{6}$ **7.** $-(7+2)$ **8.** 0

9. π **10.** $-\sqrt{5}$ **11.** x **12.** $-a$

Since all real numbers are classified as positive, negative, or zero, this set can be represented on a number line. By using a straight line from plane geometry and choosing a point to be represented by 0 and another point to be represented by 1, we can, with certain agreements, establish a correspondence between the real numbers and the points of the line. We must first agree to place the positive numbers to the right of 0 and the negative numbers to the left of 0. We must also agree that the length of the line segment from 0 to 1 will be used as a unit measurement between all real numbers that differ by 1, and that real numbers between 0 and 1 will be represented by proportional parts of this unit. Such a **real number line** is represented below.

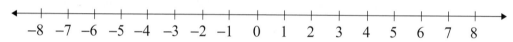

The arrows indicate that the line extends infinitely in each direction.

Example 3   We can illustrate +3 by starting at zero on the number line and moving three units to the right.

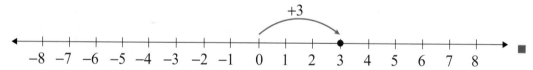

Example 4   We can illustrate –2 by starting at zero on the number line and moving two units to the left.

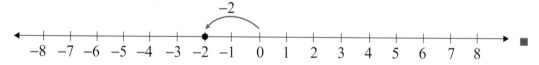

When we want to determine the distance of a number from zero without regard to direction, we are finding the **absolute value** of the number. The symbol $|x|$ is read "the absolute value of x." We can think of $|x|$ as the distance of the real number x from zero on the number line. Distance is always nonnegative.

Example 5 $|5| = 5$ since 5 is five units from zero on the number line. ■

Example 6 $|-7| = 7$ since –7 is seven units from zero on the number line. ■

These examples lead us to the following definition of **absolute value**.

$$|x| = x \qquad \text{if } x \text{ is 0 or to the right of 0.}$$

$$|x| = -x \qquad \text{if } x \text{ is to the left of 0.}$$

Notice this definition states that the absolute value of a number can never be negative. For instance, $|4| = 4$ because 4 is to the right of 0, and $|-4| = -(-4) = 4$ because -4 is to the left of 0.

EXERCISE 1-1-3

Find the following absolute values.

1. $|2|$

2. $\left|\dfrac{8}{5}\right|$

3. $|-5|$

4. $\left|-\dfrac{3}{5}\right|$

5. $|0|$

6. $-|-2|$

7. $|-\pi|$

8. $\left|\sqrt{5}\right|$

The operation of addition can be thought of as finding the end result when numbers are combined. We can illustrate addition of signed numbers by using the number line if we think of a positive number as movement along the line to the right and a negative number as movement along the line to the left.

Example 7   Use the number line to find the sum of 3 and 4.

Solution   Starting at 0 on the number line, we move three units to the right. From the point 3 we then move four units to the right.

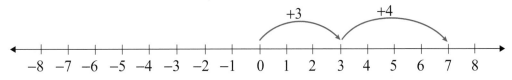

Notice that the result of these two movements is 7. Therefore, we can say that $3 + 4 = 7$. ∎

Example 8   Find: $(-3) + (-4)$

Solution   Starting at 0 on the number line, we move three units to the left. From the point -3 we then move four units to the left.

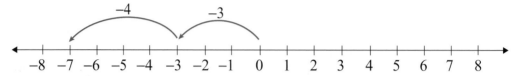

Since the end result is -7, we have $(-3) + (-4) = -7$. ∎

If we move in a certain direction along the line and then move even further in that same direction, the end result will be the combination of the two movements in that same direction. We state this fact in a formal rule.

> **RULE**
>
> To add real numbers having like signs, add the absolute values of the numbers and use the common sign.

Example 9 $12 + 5 = 17$ ∎

Example 10 $-8 + (-7) = -15$ ∎

Example 11 $-3 + (-6) = -9$ ∎

Example 12 $-\dfrac{1}{2} + \left(-\dfrac{3}{4}\right) = -\dfrac{2}{4} + \left(-\dfrac{3}{4}\right) = -\dfrac{5}{4}$ ∎

Note: To add fractions, you must have a common denominator. You combine the numerators over the common denominator.

We now look at the other possibility, adding signed numbers when the signs are not alike.

Example 13   Add 7 and -2 using the number line.

Solution   Starting at 0, we move seven units to the right to the point 7. From this point we then move two units to the left.

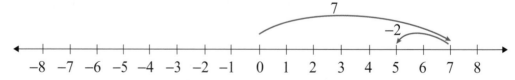

We see the end result is 5, so we conclude $7 + (-2) = 5$. Notice that we could have illustrated this problem by first moving to the left two units and then to the right seven units.

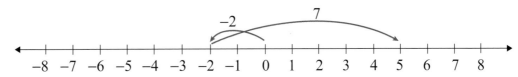

Our end result is still 5. ■

Example 13 illustrates the *commutative property of addition* which is listed as Property 2 on the front cover.

$$7 + (-2) = (-2) + 7$$

Example 14   Use the number line to find $-8 + 3$.

Solution   Starting at 0, we move eight units to the left to the point -8. From there we move three units to the right.

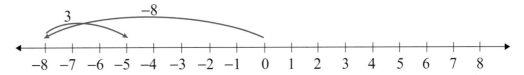

The end result is -5, so $-8 + 3 = -5$. ■

When we start at 0 and move along the number line in one direction and then in the opposite direction, and the greatest movement is to the right, the end result will be to the right of zero. Conversely, when the greatest movement is to the left, the end result will be to the left of zero. We state this formally as a rule.

RULE

To add real numbers with unlike signs, subtract the absolute values of the numbers. The sign of the number with the larger absolute value is the sign of the answer.

Example 15   Add: $6 + (-5)$

Solution   Subtract the absolute values: $6 - 5 = 1$. Since 6 is larger than 5, the answer is positive.
$6 + (-5) = 1$. ■

Example 16   Add: $-6 + 5$

Solution   Subtract the absolute values: $6 - 5 = 1$. Since 6 is larger than 5, the answer is negative.
$-6 + 5 = -1$. ■

Example 17   Add: $-50 + 17$

Solution   Subtract the absolute values: $50 - 17 = 33$. Since 50 is larger than 17, the answer is negative.
$-50 + 17 = -33$. ■

Example 18 Add: $\dfrac{7}{8} + \left(-\dfrac{2}{8} \right)$

Solution Subtract the absolute values: $\dfrac{7}{8} - \dfrac{2}{8} = \dfrac{5}{8}$. Since $\dfrac{7}{8}$ is larger than $\dfrac{2}{8}$, the answer is positive.

$$\dfrac{7}{8} + \left(-\dfrac{2}{8} \right) = \dfrac{5}{8}. \ \blacksquare$$

The following is an application of adding signed numbers.

Example 19 From 6:00 P.M. until midnight the temperature falls four degrees, and from midnight until 4:00 A.M. it falls eight more degrees. What is the total fall in temperature from 6:00 P.M. to 4:00 A.M.?

Solution If a fall in temperature is represented by a minus sign, then a fall of $4°$ followed by a fall of $8°$ can be written as $\left(-4° \right) + \left(-8° \right) = -12°$. $\blacksquare$

EXERCISE 1-1-4

Find the sums.

1. $7 + 10$

2. $16 + 28$

3. $(-6) + (-9)$

4. $(-3) + (-21)$

5. $-11 + 33$

6. $-13 + 27$

7. $1 + (-5)$

8. $13 + (-15)$

9. $-35 + 35$

10. $22 + (-22)$

11. $25 + 15$

12. $-13 + (-26)$

13. $\dfrac{3}{7} + \left(-\dfrac{5}{7}\right)$

14. $\left(-\dfrac{2}{5}\right) + \dfrac{4}{5}$

15. $\dfrac{1}{2} + \left(-\dfrac{1}{2}\right)$

16. $\pi + (-\pi)$

17. $\dfrac{2}{3} + \left(-\dfrac{1}{2}\right)$

18. $-\dfrac{5}{4} + \dfrac{1}{3}$

19. Find the result of a rise in temperature of eight degrees followed by a fall of two degrees.

20. Find the result of a loss of $36 combined with a profit of $30.

21. The stock market gained ten points in the morning, then lost six points in the afternoon. What was the net gain or loss for the day?

22. A worker earned $25 in the morning and spent $18 in the afternoon. What was his net gain or loss for the day?

23. Find the result of a gain of two yards in a football game followed by a loss of six yards.

1–2   SUBTRACTION OF REAL NUMBERS

> **OBJECTIVES**
>
> Upon completion of this section you should be able to:
> 1. Subtract one real number from another.
> 2. Add and subtract several real numbers.

Now that we have learned to add real numbers, the next basic operation we will study is subtraction. **Subtraction** is defined as adding the opposite of the number being subtracted. The result is called the **difference**.

> **RULE**
>
> To subtract one signed number from another, change the sign of the number being subtracted and use the appropriate rule for addition.

Example 1   Subtract: $5 - 2$

Solution   Change 2 to –2 and add. $5 + (-2) = 3$
So $5 - 2 = 3$. ∎

Example 2   Subtract: $5 - (-2)$

Solution   Change –2 to 2 and add. $5 + 2 = 7$
So $5 - (-2) = 7$. ∎

Example 3   Subtract: $-8 - 10$

Solution   Change 10 to –10 and add. $-8 + (-10) = -18$
So $-8 - 10 = -18$. ∎

Example 4   Subtract: $-8 - (-10)$

Solution   Change –10 to 10 and add. $-8 + 10 = 2$
So $-8 - (-10) = 2$. ∎

Example 5   Subtract: $-\dfrac{1}{5} - \dfrac{2}{3}$

Solution   Change $\dfrac{2}{3}$ to $-\dfrac{2}{3}$ and add. Remember to rewrite the fractions so that they contain a common denominator. $-\dfrac{1}{5} + \left(-\dfrac{2}{3}\right) = -\dfrac{3}{15} + \left(-\dfrac{10}{15}\right) = -\dfrac{13}{15}$
So $-\dfrac{1}{5} - \dfrac{2}{3} = -\dfrac{13}{15}$. ∎

Note: In algebra no two signs are ever written together without the use of parentheses. Thus $5 + -3$ would not be correct, but $5 + (-3)$ would be correct.

EXERCISE 1-2-1

Find the differences.

1. $9 - 3$

2. $8 - 4$

3. $-5 - 3$

4. $-7 - 6$

5. $-11 - (-21)$

6. $-13 - (-53)$

7. $8 - (-9)$

8. $21 - (-13)$

9. $-14 - 9$

10. $-21 - 2$

11. $25 - 36$

12. $17 - 54$

13. $-12 - 15$

14. $32 - (-5)$

15. $-14 - (-14)$

16. $\left(-\dfrac{3}{5}\right) - \left(-\dfrac{3}{5}\right)$

17. $\left(-\dfrac{5}{9}\right) - \dfrac{1}{3}$

18. $\dfrac{5}{8} - \left(-\dfrac{1}{4}\right)$

19. $-\dfrac{3}{8} - \dfrac{1}{4}$

20. $\dfrac{10}{9} - \dfrac{5}{3}$

We have established the rules for adding and subtracting signed numbers. We now need to look at expressions that involve both operations, such as

$$3 - 2 + 5.$$

If one person should decide to subtract 2 from 3 and add the result to 5, the answer would be 6. If another person should decide to add 2 and 5 and subtract the result from 3, the answer would then be –4. Which is correct? To determine the correct answer, let's change the subtraction to addition and add.

$$3 + (-2) + 5$$

Using a combination of the commutative and associative properties of addition (Properties 3 and 4 on the front cover), we can now add these numbers in any order.

$$3 + (-2) + 5 \qquad 3 + (-2) + 5 \qquad 3 + (-2) + 5$$
$$1 + 5 \quad \text{or} \quad 3 + 3 \quad \text{or} \quad 8 + (-2)$$
$$6 \qquad\qquad 6 \qquad\qquad 6$$

So 6 is the correct answer. This is the answer the first person got by performing the operations in order from left to right.

> **RULE**
>
> To simplify an expression that contains only additions and subtractions, perform these operations in order from left to right,
>   OR
> rewrite all subtractions as adding the opposite and then add.

Example 6   Evaluate: $-3 - 4 + 9 - 1$

Solution   METHOD 1:

Subtract. $-3 + (-4) + 9 - 1$

$-7 + 9 - 1$

Add. $2 - 1$

Subtract. 1

METHOD 2:

Change to addition. $-3 + (-4) + 9 + (-1)$

Add the numbers

with like signs. $-3 + (-4) + (-1) + 9$

$-8 + 9$

1 ■

Example 7   Evaluate: $7 - 2 + 3 - 5 + 4$

Solution   METHOD 1:

$5 + 3 - 5 + 4$

$8 - 5 + 4$

$3 + 4$

7

METHOD 2:

$7 + (-2) + 3 + (-5) + 4$

$7 + 3 + 4 + (-2) + (-5)$

$14 + (-7)$

7 ■

Example 8   At 9:00 A.M. the temperature is $20°$. From 9:00 A.M. until 1:00 P.M. it rises $14°$, then from 1:00 P.M. until 5:00 P.M. it falls $18°$. What is the temperature at 5:00 P.M.?

Solution   We can represent a rise in temperature as addition and a fall in temperature as subtraction. So we write this example as $20° + 14° - 18°$.

$$20° + 14° - 18° = 34° - 18° = 16°$$

The temperature at 5:00 P.M. is $16°$. ■

EXERCISE 1-2-2

Evaluate.

1. $3 + 5 + 7$

2. $16 + 4 + 2$

3. $-5 - 13 - 7$

4. $-7 - 5 - 13$

5. $5 - 6 + 4$

6. $9 - 7 - 6$

7. $-3 - 7 - 5$

8. $4 - 7 - (-6)$

9. $\dfrac{4}{5} + \dfrac{7}{5} - \dfrac{13}{5}$

10. $\dfrac{7}{3} - \dfrac{4}{3} - \dfrac{2}{3}$

11. $13 - 17 + 5 - 4$

12. $-4 + 5 - 12 + 6$

13. $-7 - (-5) + 7 - (-9)$

14. $-15 - (-37) - 15 + 3$

15. $\dfrac{1}{2} - \dfrac{3}{4} - \dfrac{5}{8} + \dfrac{3}{2}$

16. $\dfrac{2}{3} - \dfrac{7}{6} - \dfrac{1}{2} - \dfrac{1}{3}$

17. $8 - 7 + 2 - 4 + 9$

18. $-19 + 7 - 22 + 16$

19. $34 - 16 + 9 + 16 - 3$ **20.** $9 - 5 + 13 + 6 - 7 - 20$

21. If the temperature was $50°$ at 6:00 A.M. and from 6:00 A.M. to 2:00 P.M. it rose $15°$, then from 2:00 P.M. to 7:00 P.M. it fell $9°$, what was the temperature at 7:00 P.M.?

22. A temperature starts at $10°$. It falls $15°$, then rises $8°$, then falls $3°$. What is the final temperature?

23. A dieter lost five pounds the first week, gained two pounds the second week, lost three pounds the third week, and lost two pounds the fourth week. What was the dieter's net gain or loss for the four weeks?

24. In American football a team has four plays (downs) during which the ball must be moved ten yards forward or given up to the other team. On a series of four downs the Pittsburgh Steelers did the following:

first down   gained five yards
second down   lost eight yards
third down   gained twelve yards
fourth down   gained two yards

How many yards did they gain in the four downs? Did the team gain enough yards to keep the ball?

1–3   MULTIPLICATION AND DIVISION OF REAL NUMBERS

> ### OBJECTIVES
>
> Upon completion of this section you should be able to:
> 1. Multiply and divide real numbers.
> 2. Find the product of several real numbers.

We will now study the two remaining basic operations, multiplication and division. The result of multiplication is called a **product** and the numbers to be multiplied are called **factors**. There are several ways to indicate the product of two numbers. For example, 2×3, $2 \cdot 3$, $(2) \times (3)$, $(2)(3)$, $(2) \cdot (3)$, and $2(3)$ are all ways of indicating the product of the factors 2 and 3. If letters are used to represent numbers, multiplication can be indicated by writing the letters together without a sign or parentheses. For instance, if x and y represent numbers, then xy represents the product of x and y.

To multiply fractions, multiply the numerators and then multiply the denominators: $\left(\dfrac{a}{b}\right)\left(\dfrac{c}{d}\right) = \dfrac{ac}{bd}$.

The result of division is called a **quotient**. The number we divide by is called the **divisor**. To indicate a quotient, we write $x \div y$ or $\dfrac{x}{y}$. Division can be defined as multiplying by the **reciprocal** of the divisor. (For the definition of reciprocal, see Property 10 on the front cover.) So dividing by 2 is the same as multiplying by $\dfrac{1}{2}$ and dividing by $\dfrac{3}{4}$ is the same as multiplying by $\dfrac{4}{3}$. To divide fractions, multiply the first fraction by the reciprocal of the divisor: $\dfrac{a}{b} \div \dfrac{c}{d} = \left(\dfrac{a}{b}\right)\left(\dfrac{d}{c}\right) = \dfrac{ad}{bc}$. Since division can be rewritten as multiplication, any rules we define for multiplication of signed numbers will also apply to division of signed numbers.

The rules for determining the sign of a product or quotient of signed numbers are not the same as those for determining the sign of a sum of signed numbers.

> ### RULE
>
> 1. The product or quotient of two numbers having like signs will be positive.
> 2. The product or quotient of two numbers having unlike signs will be negative.

To illustrate this rule, we think of multiplication as repeated addition. Thus, $3(5) = 5 + 5 + 5 = 15$. This is an example of positive times positive is positive (like signs $\rightarrow$ positive result). Also, $4(-3) = (-3) + (-3) + (-3) + (-3) = -12$. This is an example of positive times negative is negative (unlike signs $\rightarrow$ negative result).

To illustrate negative times negative is positive, let us find the product $(-3)(-5)$. We will make use of three properties: additive inverse (Property 5 on the front cover), distributive (Property 11 on the front cover), and the property that the product of zero and any real number is 0 (Zero Factor Property).

Consider the expression $(-3)(5) + (-3)(-5).$

Use the distributive property to rewrite $(-3)(5) + (-3)(-5) = (-3)(5 + (-5)).$

The additive inverse gives $(-3)(5 + (-5)) = (-3)(0).$

The zero factor property gives $(-3)(0) = 0.$

Since $-15 + 15 = 0$ and $(-3)(5) + (-3)(-5) = -15 + (-3)(-5) = 0$, we can conclude that $(-3)(-5) = 15.$

Example 1 $(3)(2) = 6$ $(+)(+) = +$ ■

Example 2 $(-3)(-2) = 6$ $(-)(-) = +$ ■

Example 3 $(3)(-2) = -6$ $(+)(-) = -$ ■

Example 4 $(-3)(2) = -6$ $(-)(+) = -$ ■

Example 5 $\dfrac{6}{2} = 3$ $\dfrac{+}{+} = +$ ■

Example 6 $\dfrac{-6}{-2} = 3$ $\dfrac{-}{-} = +$ ■

Example 7 $\dfrac{6}{-2} = -3$ $\dfrac{+}{-} = -$ ■

Example 8 $\dfrac{-6}{2} = -3$ $\dfrac{-}{+} = -$ ■

Example 9 $(-10) \div (-2) = 5$ ■

Example 10 $\left(\dfrac{2}{3}\right)\left(\dfrac{5}{7}\right) = \dfrac{10}{21}$ ■

Example 11 $(-2)\left(-\dfrac{1}{5}\right) = \dfrac{2}{5}$ ■

Example 12 $10 \div (-2) = -5$ ■

Example 13 $\left(-\dfrac{1}{6}\right)\left(\dfrac{2}{7}\right) = -\dfrac{2}{42} = -\dfrac{1}{21}$ ■

Example 14 $7\left(-\dfrac{3}{5}\right) = -\dfrac{21}{5}$ ■

CAUTION   Do not confuse the rules of signs for addition and subtraction with the rules for multiplication and division. All work from this point on will be easier if you pause now and make sure that you know the rules of signs.

Note: When working the following exercises, remember that 0 divided by a nonzero real number is 0, and division by 0 is undefined.

EXERCISE 1-3-1

Find the products and quotients.

1. $(-3)(-4)$

2. $(-3)(4)$

3. $5(-2)$

4. $-7(3)$

5. $(6)(2)$

6. $(-7)\left(-\dfrac{1}{7}\right)$

7. $\left(\dfrac{3}{4}\right)\left(\dfrac{4}{3}\right)$

8. $\left(\dfrac{2}{3}\right)(-12)$

9. $\left(-\dfrac{2}{5}\right) \div \left(\dfrac{3}{7}\right)$

10. $\left(-\dfrac{1}{2}\right) \div \left(-\dfrac{1}{8}\right)$

11. $\dfrac{-15}{-3}$

12. $\dfrac{18}{-6}$

13. $20 \div (-5)$

14. $20 \div 5$

15. $\dfrac{0}{-6}$

16. $(-6)\left(\dfrac{2}{3}\right)$

17. $\dfrac{(-6)}{\frac{3}{2}}$

18. $(-10)\left(\dfrac{3}{4}\right)$

19. $\dfrac{-2}{0}$

20. $\left(-\dfrac{7}{8}\right)\left(-\dfrac{4}{3}\right)$

Multiplication, like addition and subtraction, can be performed on only two numbers at a time. If the product of three numbers is indicated, the product of any two of them must be multiplied by the third. It does not matter which numbers you choose to multiply first since multiplication is both commutative and associative (See Properties 7 and 8 on the front cover).

Example 15   Multiply: $(-2)(-3)(4)$

Solution $(-2)(-3)(4)$ or $(-2)(-3)(4)$ or $(-2)(-3)(4)$
                  $(6)(4)$ $(-2)(-12)$ $(-8)(-3)$
                  24   24   24 ∎

If the product of more than three numbers is indicated, this procedure is simply repeated.

Example 16   Multiply: $(-2)(3)(-4)(5)$

Solution   One way to do this is as follows.

$$(-2)(3)(-4)(5)$$
$$(-6)(-4)(5)$$
$$(-6)(-20)$$
$$120 \ ∎$$

Example 17   Multiply: $(-3)(2)\left(\dfrac{-1}{5}\right)(1)(3)$

Solution $(-6)\left(\dfrac{-1}{5}\right)(3)$

$$\left(\dfrac{6}{5}\right)(3)$$

$$\dfrac{18}{5} \quad ∎$$

EXERCISE 1-3-2

Find the products.

1. $(2)(5)(-7)$ **2.** $(4)(-6)(3)$ **3.** $(10)(1)(-8)$ **4.** $(-6)(2)(-3)$

5. $(-4)(-6)(2)$ **6.** $(6)(-2)(-5)$ **7.** $(-4)(-9)(-3)$ **8.** $(-2)(-8)(-5)$

9. $(-1)(-5)\left(-\dfrac{1}{5}\right)$ **10.** $\left(-\dfrac{1}{2}\right)\left(-\dfrac{4}{5}\right)\left(-\dfrac{3}{2}\right)$ **11.** $(-7)(-14)(0)$

12. $(3)(-1)(0)(10)$ **13.** $(-4)(3)(-11)(2)$ **14.** $\left(-\dfrac{8}{9}\right)\left(\dfrac{1}{4}\right)\left(-\dfrac{3}{5}\right)(10)$

15. $\left(\dfrac{3}{5}\right)\left(-\dfrac{2}{3}\right)(-1)\left(\dfrac{7}{4}\right)$ **16.** $(-8)(-5)(-2)(-3)$ **17.** $(-2)(-1)(-10)(4)(1)$

18. $\left(-\dfrac{1}{2}\right)(-14)\left(-\dfrac{3}{7}\right)(-2)$ **19.** $(27)\left(-\dfrac{1}{9}\right)\left(-\dfrac{1}{3}\right)(-37)$ **20.** $(49)(-72)(-104)(0)(-23)$

Practice your skills.

21. $-(3+2)$ **22.** $\left|-\dfrac{2}{3}\right|$ **23.** $(-3)\left(\dfrac{3}{5}\right)$ **24.** $-2-7$

25. $-\dfrac{1}{2}+\dfrac{3}{4}$ **26.** $(-4)(2)(5)(-1)$ **27.** $-1+6-4+10$ **28.** $|-5|+3$

1-4 POSITIVE WHOLE NUMBER EXPONENTS

OBJECTIVES

Upon completion of this section you should be able to:

1. Evaluate expressions such as 2^4.
2. Find the products and quotients of expressions involving exponents.

In algebra, letters of the alphabet, both uppercase and lowercase, are used to represent numbers as we perform various operations. These letters are called **variables**. Special mathematical symbols and notations are used to simplify expressions. One such symbol is the *exponent.*

Recall that numbers multiplied together are called **factors**. When a factor is repeated, we use a shorthand notation. In the expression 3^2, the number 2 is called an **exponent** and the number 3 is called the **base**. The exponent indicates that the base is used as a factor two times. Thus 3^2 means 3(3). Likewise, x^2 indicates that x is used as a factor two times.

$$x^2 = (x)(x)$$
$$x^3 = (x)(x)(x)$$
$$x^5 = (x)(x)(x)(x)(x)$$

The expression x^2 is read "x squared" or "x to the second power." The expression x^3 is read "x cubed" or "x to the third power." When the exponent is four or more, then x^n is read only "x to the nth power." For example, x^5 is read "x to the fifth power." If no exponent is written, it is understood to be 1.

$$x^1 = x$$

Example 1   Evaluate: 5^2

Solution   The exponent 2 means to use the base 5 as a factor 2 times.

$$5^2 = 5(5) = 25 \quad \blacksquare$$

Example 2   Evaluate: $(-3)^3$

Solution   The exponent 3 means to use the base -3 as a factor 3 times.

$$(-3)(-3)(-3) = -27 \quad \blacksquare$$

Example 3   Evaluate: $-(2)^4$

Solution   We are asked to find the opposite or negative of 2^4.

$-(2)(2)(2)(2) = -16$ ∎

Note: $-(2)^4$, $-(2^4)$, and -2^4 are all ways of expressing the opposite of 2^4.

Example 4   Evaluate: $(-2)^4$

Solution $(-2)(-2)(-2)(-2) = 16$ ∎

From the definition of an exponent, we can establish laws for multiplying and dividing numbers with exponents.

For instance, suppose we wish to multiply $x^3 \cdot x^5$. From the definition we have

$$x^3 = x \cdot x \cdot x \text{ and } x^5 = x \cdot x \cdot x \cdot x \cdot x$$
$$\text{so } x^3 \cdot x^5 = (x \cdot x \cdot x) \cdot (x \cdot x \cdot x \cdot x \cdot x)$$
$$= x \cdot x \cdot x \cdot x \cdot x \cdot x \cdot x \cdot x$$
$$= x^8.$$

If we generalize the above example to the product of $x^a \cdot x^b$, we have

$$x^a = \underbrace{x \cdot x \cdot x \cdot \ldots x}_{a \text{ factors}}$$

$$x^b = \underbrace{x \cdot x \cdot x \cdot \ldots x}_{b \text{ factors}}$$

$$\text{and } x^a \cdot x^b = \underbrace{(x \cdot x \cdot x \cdot \ldots x)}_{a \text{ factors}}\underbrace{(x \cdot x \cdot x \cdot \ldots x)}_{b \text{ factors}}$$

$$\text{so } x^a \cdot x^b = x^{a+b}.$$

This is the first law of exponents.

FIRST LAW OF EXPONENTS

$x^a \cdot x^b = x^{a+b}$ (To multiply factors with the same base, keep the base and add the exponents.)

CAUTION   Be careful to note that the rule just stated applies only when the bases are the same. Also note that only the exponents are added. The base never changes.

Example 5   Find the product: $x^3 \cdot x^4$

Solution   Since both factors have the same base, x, we keep x as the base and add 3 and 4.
$x^3 \cdot x^4 = x^{3+4} = x^7$ ∎

Example 6   Find the product: $y^8 \cdot y$

Solution   Since both factors have the same base, y, we keep y as the base and add 8 and 1.
$y^8 \cdot y = y^{8+1} = y^9$ ∎

Example 7   Find the product: $5^4 \cdot 5^7$

Solution   Since both factors have the same base, 5, we keep 5 as the base and add 4 and 7.
$5^4 \cdot 5^7 = 5^{4+7} = 5^{11}$ ∎

Example 8   Find the product: $x^3 \cdot y^2$

Solution   The law does not apply because the bases are not the same, so this is written as $x^3 y^2$. ∎

Example 9 $(-2)^3 (-2)^4 = (-2)^7$ ∎

Example 10 $(a^3)(a^3)(a) = a^7$ ∎

Example 11 $3^2 \cdot 2^4$ The law does not apply. ∎

EXERCISE 1-4-1

Evaluate.

1. 3^2

2. $(4)^3$

3. $(-3)^2$

4. $(-2)^2$

5. -3^2

6. -5^3

7. $-(2)^3$

8. $-(-2)^3$

Find the following products. Give your answers in exponential form.

9. $x^3 \cdot x^5$

10. $y^4 \cdot y^2$

11. $3^4 \cdot 3^3$

12. $2^3 \cdot 2^5$

13. $(-5)(-5)^3$

14. $(-3)^2(-3)^7$

15. $(x)(x^3)(x^4)$

16. $(a^3)(a)(a^2)$

17. $(x^4)(y^3)$

18. $(2^3)(a^4)$

We now will consider expressions that contain both numbers and letters.

In an expression such as $3x^4$, which means $3(x)(x)(x)(x)$, 3 is called the **coefficient**, x is called the **base**, and 4 is called the **exponent**.

coefficient $\longrightarrow 3x^4 \longleftarrow$ exponent

base

Note that only the x and *not* the 3 is raised to the 4th power. If both the 3 and the x were raised to the 4th power, the expression would be written as $(3x)^4$. Notice the parentheses!

$$(3x)^4 = (3x)(3x)(3x)(3x)$$

Also note that if no coefficient is present, it is understood to be 1.

An indicated product such as $(2x^3)(3x^4)$ involves two previously discussed ideas. The coefficients, 2 and 3, are multiplied since the order of multiplication has no effect on the answer. The factors, x^3 and x^4, can also be multiplied by keeping the base and adding the exponents since the bases are the same.

$$(2x^3)(3x^4)$$
$$(2)(3)(x^3)(x^4)$$
$$6x^7$$

Example 12   Simplify: $\left(5x^2\right)\left(-2x^3\right)$

Solution   Multiply the coefficients, 5 and -2. Multiply x^2 and x^3.

$$(5)(-2)\left(x^2\right)\left(x^3\right)$$
$$-10x^5 \quad \blacksquare$$

Example 13   Simplify: $\left(3x^2\right)\left(2y^2\right)\left(4z^5\right)$

Solution   Multiply the coefficients. The bases are not the same, so you cannot add the exponents.

$$(3)(2)(4)\left(x^2\right)\left(y^2\right)\left(z^5\right)$$
$$24x^2y^2z^5 \quad \blacksquare$$

EXERCISE 1-4-2

Simplify.

1. $\left(2x^2\right)\left(5x^3\right)$

2. $\left(4x^3\right)\left(3x^5\right)$

3. $\left(-3x^2\right)\left(7x^3\right)$

4. $\left(11a^3\right)\left(-6a^7\right)$

5. $\left(-\dfrac{8}{3}a^2\right)\left(-\dfrac{3}{4}b^3\right)$

6. $\left(\dfrac{7}{2}xy\right)\left(\dfrac{2}{5}xy\right)$

7. $\left(-2x^2y\right)\left(3xy^3\right)$

8. $\left(6a^2b\right)\left(5a^3c\right)$

9. $\left(\dfrac{1}{2}x^2\right)\left(-\dfrac{2}{5}x^5\right)\left(3x^4\right)$

10. $\left(-\dfrac{7}{2}a^2b^2\right)\left(-2ab\right)\left(\dfrac{3}{7}ab\right)$

11. $\left(-6xy\right)\left(-2x^2\right)\left(-4y^5\right)$

12. $\left(11ab\right)\left(3b^2c\right)\left(-5a^2c^3\right)$

13. $\left(-\dfrac{3}{2}x^2y^3\right)\left(-\dfrac{4}{7}xy^5z^2\right)\left(\dfrac{5}{3}yz\right)$

14. $\left(6x^2\right)\left(-5y^2\right)(2z)$

15. $\left(-2a^2b\right)\left(-5b^2c^2\right)\left(6c^3d^2\right)$

Now we will establish a rule for dividing expressions with exponents. For instance, suppose we wish to divide $\dfrac{x^5}{x^3}$. From the definition of exponents we have $\dfrac{x^5}{x^3} = \dfrac{(x)(x)(x)(x)(x)}{(x)(x)(x)}$. We can now use a rule from arithmetic that states a nonzero number divided by itself is equal to 1. Thus, assuming $x \neq 0$, we may divide three x's in the numerator by the three x's in the denominator

obtaining $\qquad \dfrac{x^5}{x^3} = \dfrac{(\cancel{x})(\cancel{x})(\cancel{x})(x)(x)}{(\cancel{x})(\cancel{x})(\cancel{x})} = \dfrac{x^2}{1} = x^2.$

Now consider $\dfrac{x^3}{x^5}$. Again from the definition of exponents and assuming $x \neq 0$, we have $\qquad \dfrac{x^3}{x^5} = \dfrac{(\cancel{x})(\cancel{x})(\cancel{x})}{(\cancel{x})(\cancel{x})(\cancel{x})(x)(x)} = \dfrac{1}{x^2}.$

SECOND LAW OF EXPONENTS

If $x \neq 0$,

$$\frac{x^a}{x^b} = x^{a-b} \text{ if } a \text{ is greater than } b;$$

$$\frac{x^a}{x^b} = \frac{1}{x^{b-a}} \text{ if } a \text{ is less than } b;$$

$$\frac{x^a}{x^b} = 1 \text{ if } a = b.$$

Example 14   Simplify: $\dfrac{x^5}{x^3}$

Solution   Since the numerator and denominator have like bases and the larger exponent is in the numerator, we keep the base, x, in the numerator and subtract $5-3$.

$$\dfrac{x^5}{x^3} = \dfrac{x^{5-3}}{1} = x^2 \quad \blacksquare$$

Example 15   Simplify: $\dfrac{x^3}{x^5}$

Solution   Since the numerator and denominator have like bases and the larger exponent is in the denominator, we keep the base, x, in the denominator and subtract $5-3$.

$$\dfrac{x^3}{x^5} = \dfrac{1}{x^{5-3}} = \dfrac{1}{x^2} \quad \blacksquare$$

Example 16 $\dfrac{x^{10}}{x^7} = x^{10-7} = x^3 \quad \blacksquare$

Example 17 $\dfrac{y^3}{y^8} = \dfrac{1}{y^{8-3}} = \dfrac{1}{y^5} \quad \blacksquare$

Example 18 $\dfrac{a^3}{a^3} = 1 \quad \blacksquare$

Example 19 $\dfrac{2^4}{2^6} = \dfrac{1}{2^{6-4}} = \dfrac{1}{2^2} = \dfrac{1}{4} \quad \blacksquare$

Example 20 $\dfrac{x^3}{y^2}$ This expression is in simplest form. $\blacksquare$

Example 21   Simplify: $\dfrac{-6x^5}{8x^3}$

Solution   The coefficients are treated as a fraction to be simplified.

$$\dfrac{-6x^5}{8x^3} = \dfrac{-3x^{5-3}}{4} = \dfrac{-3x^2}{4} \quad \blacksquare$$

Example 22 $\dfrac{15a^2b^7}{-3a^5b^3} = \dfrac{-5b^{7-3}}{a^{5-2}} = \dfrac{-5b^4}{a^3} \quad \blacksquare$

> **RULE**
>
> Only similar terms can be added or subtracted. To add or subtract similar terms, combine the coefficients and use this result as the coefficient of the common variable factors. (This is an application of the distributive property.)

Example 8   Simplify: $2xy + 3xy$

Solution   Since $2xy$ and $3xy$ are similar, we add the coefficients 2 and 3. The result, 5, is the coefficient of the common variable factors.

$$(2+3)xy$$
$$5xy \quad \blacksquare$$

Example 9   Simplify: $2x^2 + 3x^3$

Solution   These terms cannot be combined because x^2 and x^3 are not the same. This expression is already in simplest form. ∎

Example 10   Simplify: $5x^2y - 3x^2y$

Solution $$(5-3)x^2y$$
$$2x^2y \quad \blacksquare$$

Example 11   Simplify: $-7xz - 2xz$

Solution $$(-7-2)xz$$
$$-9xz \quad \blacksquare$$

Example 12   Simplify: $-3xy + 2x$

Solution   This expression is already in simplest form. ∎

Example 13   Simplify: $2xy + 3xy - 4x$

Solution   This expression is simplified by combining the two terms that are similar.

$$2xy + 3xy - 4x$$
$$5xy - 4x \quad \blacksquare$$

To simplify variable expressions containing addition and subtraction, all similar terms are combined. If more than one term occurs in the simplified form, the order of the terms is not important. The answer for Example 13 could be written as $5xy - 4x$ or $-4x + 5xy$.

EXERCISE 1-5-1

Simplify by combining similar terms.

1. $5a + 7a$

2. $9x - 5x$

3. $3a - 10a$

4. $11x^2 + 7x^2$

5. $8xy - 12xy$

6. $15x - 5y$

7. $17x^3 + 3x^2$

8. $13xyz + 9xyz$

9. $-7ab^2c - 5abc$

10. $x^2y + 12x^2y$

11. $2x - 7x + 5x$

12. $20a + 13b - 33a$

13. $16a^2b - 31a^2b + 5ab^2$

14. $8x^2y - 3x^2y + 5x^2y$

15. $2ab + 5bc - 6ac$

16. $5xy^2 - xy^2 + 2x^2y^2$

17. $4xy + 13a - 7xy - 5a$

18. $16a^2b - 4a^2b + 5a^2b^2 - 7a^2b$

19. $21xyz + 15xy - 17xyz + 7xy^2$

20. $6xy - 5yz - 9xy + 7yz + 3xy - 2yz$

1–6   THE PRODUCT OF A MONOMIAL AND ANOTHER POLYNOMIAL

> **OBJECTIVES**
>
> Upon completion of this section you should be able to:
> 1. Identify monomials, binomials, and trinomials.
> 2. Multiply a polynomial by a monomial.

In Section 1–4 we learned to multiply expressions with one-term factors such as $(2x^3)(-3x^4)$. Now we will extend this process. To do so we need to use the *distributive property* of multiplication over addition, $a(b+c) = ab+ac$. This property gives us a way of changing from a product of factors, (a) and $(b+c)$, to a sum of terms, ab and ac. The distributive property can also be written $(b+c)a = ba+ca$.

The distributive property is not limited to two terms. It is true for any number of terms.

$$a(b+c+d+\ldots) = ab+ac+ad+\ldots$$

Example 1   Expand: $3(-2+4+5)$

Solution $3(-2)+3(4)+3(5) = -6+12+15$
$$= 21 \ \blacksquare$$

An algebraic expression having one or more terms that contain only whole number exponents is called a **polynomial**. Special names are sometimes used for polynomials of one, two, or three terms. A polynomial having *one* term is called a **monomial**. A polynomial with *two* terms is called a **binomial**. A polynomial with *three* terms is called a **trinomial**.

Example 2 $5xy$ is a monomial. $\blacksquare$

Example 3 $5x+y$ is a binomial. $\blacksquare$

Example 4 $3x^2y^4z$ is a monomial. $\blacksquare$

Example 5 $3x+2y-4z$ is a trinomial. $\blacksquare$

Polynomials of more than three terms generally have no special names.

> **RULE**
>
> To multiply a polynomial by a monomial, multiply *each term* of the polynomial by the monomial. (Note that this is an application of the distributive property.)

Example 6   Multiply: $5x(2x-3y)$

Solution   First rewrite using the distributive property. Then multiply.

$$(5x)(2x)-(5x)(3y)$$
$$10x^2-15xy \quad \blacksquare$$

Example 7   Multiply: $-3(2x^2-5x+7)$

Solution
$$(-3)(2x^2)-(-3)(5x)+(-3)(7)$$
$$-6x^2-(-15x)+(-21)$$
$$-6x^2+15x-21 \quad \blacksquare$$

Note that the sign of every term of the polynomial changes when multiplying by a *negative* number.

Example 8   Multiply: $2xyz(3x^2+3y^2-3z^2)$

Solution
$$(2xyz)(3x^2)+(2xyz)(3y^2)-(2xyz)(3z^2)$$
$$6x^3yz+6xy^3z-6xyz^3 \quad \blacksquare$$

The opposite of a polynomial can be thought of as the product of -1 and the polynomial. So
$$-(a+b)=(-1)(a+b)$$
$$=(-1)a+(-1)b$$
$$=-a-b$$

Example 9   Simplify: $-(3x^2-2x+y)$

Solution
$$(-1)(3x^2-2x+y)$$
$$(-1)(3x^2)-(-1)(2x)+(-1)(y)$$
$$-3x^2-(-2x)+(-y)$$
$$-3x^2+2x-y \quad \blacksquare$$

Note that when removing parentheses preceded by a negative sign, the sign of every term within the parentheses is changed.

EXERCISE 1-6-1

Multiply.

1. $3(x+2y)$ **2.** $-2(2x+y)$ **3.** $x(x+4)$ **4.** $3x(2x-5)$

5. $-(2a-7b)$ **6.** $2a(a-5b)$ **7.** $5(2x+y-3z)$ **8.** $-6x(x-3y)$

9. $-x(2x-4y)$ **10.** $-2x(3x-2y-z)$ **11.** $-(3m+5n-p)$

12. $3mn(2m-3n-5)$ **13.** $-7y(x-2y+3)$ **14.** $2x(-x^2+3x+1)$

15. $3x(2+5x-4x^2)$ **16.** $-5y(-4y^2-2y+1)$

17. $2xy(3x^2y-2xy^2+xy)$ **18.** $-7xy(2x^2y^3+3x^2y-5xy^2)$

19. $2xyz(x^2y-5y^2z+9xz)$ **20.** $-xyz(x^2y^2z-10xz+3y^2z^3)$

1–7   ORDER OF OPERATIONS AND GROUPING SYMBOLS

OBJECTIVES

Upon completion of this section you should be able to:
1. Evaluate numerical expressions involving several operations.
2. Simplify expressions such as $2(x+y) - \left[3x - (x+y)\right]$.

As previously discussed, if a numerical expression contains only addition and subtraction, these operations are performed, in order, from left to right. Now let us consider expressions that also contain other operations. Should $5 + 3 \cdot 2$ be $8 \cdot 2$ or $5 + 6$?

RULE

If no grouping symbols occur in an expression to be evaluated, exponents (powers) must be worked first, then multiplication and division (left to right), then addition and subtraction (left to right).

So the above expression $5 + 3 \cdot 2$ would be evaluated as $5 + 6 = 11$.

A numerical expression enclosed in grouping symbols is treated as if it were a single number. Some examples of grouping symbols are parentheses, brackets, braces, absolute value, and the fraction bar. When grouping symbols occur in an expression, the operations within the grouping symbols are performed first. Therefore, the correct order in which to perform operations in an expression is

1. operations within grouping symbols,
2. exponents (powers),
3. multiplication and division (left to right),
4. addition and subtraction (left to right).

Example 1   Evaluate: $2^3 - 3 \cdot 4 + 6$

Solution

Exponent	$8 - 3 \cdot 4 + 6$
Multiplication	$8 - 12 + 6$
Addition and subtraction	2 ■

Example 2   Evaluate: $2^3 - 3(4+6)$

Solution   The parentheses change the normal order of operations. The addition is performed first.

$$2^3 - 3(10)$$
$$8 - 3(10)$$
$$8 - 30$$
$$-22 \ \blacksquare$$

Example 3   Evaluate: $5 - 4 \cdot 12 \div 3 \cdot 2$

Solution
$$5 - 48 \div 3 \cdot 2$$
$$5 - 16 \cdot 2$$
$$5 - 32$$
$$-27 \ \blacksquare$$

Example 4   Evaluate: $\dfrac{2-7}{3+7}$

Solution $\dfrac{-5}{10} = \dfrac{-1}{2}$ or $-\dfrac{1}{2}$ $\blacksquare$

EXERCISE 1-7-1

Evaluate.

1. $4(3)+6$

2. $4(3+6)$

3. $6 \div 2 + 1$

4. $6 \div (3+6)$

5. $5 \cdot 2 - 3 \cdot 2$

6. $5(2-3) \cdot 2$

7. $12 \div 2 - 6 \div 3$

8. $12 \div (2-6) \div 3$

9. $5 - 2(4)$

10. $(5-2)4$

11. $7 + 3 \cdot 5 - 2$

12. $(7+3)(5-2)$

13. $2^2 + 3 \cdot 2$

14. $3^2 + 2(7) + 5$

15. $2 \cdot 5 + 7 \cdot 3 - 5 \cdot 2$

16. $2 \cdot 8 \div 4 \cdot 3 \div 2 - 6$

17. $\dfrac{3-8}{2 \cdot 5}$

18. $\dfrac{3-2^3}{2}$

19. $\dfrac{7-2^3}{5+3}$

20. $\dfrac{4^2}{2-3 \cdot 2^3 + 6}$

21. $16 + 4 \cdot 3 - 5^2$

22. $16 + 4(3-5)^2$

23. $15 \cdot 6 \div 3 - 2^2$

24. $15(6 \div 3 - 2)^2$

25. $7 - 3^2 + |1-5| - 2$

26. $5 \cdot 3^2 - |2-6| + 4$

27. $2 - (6+4)8 \div 4$

28. $4(3-9) \div (2+1) - 3$

29. $\dfrac{4}{5} + \dfrac{27}{25} \div \dfrac{3}{5} - \dfrac{5}{2}$

30. $\dfrac{1}{3} - \left(\dfrac{16}{3}\right)\left(\dfrac{1}{8}\right) \div \left(\dfrac{4}{5}\right) + \dfrac{1}{2}$

Sometimes the meaning of an English sentence can be changed completely by the use of a comma. For example,

"Let's feed the lions, Jim."

"Let's feed the lions Jim."

Mathematical sentences may also need punctuation to clarify the meanings. Some of the symbols for punctuation or grouping of numbers are parentheses ( ), brackets [ ], and braces { }. These three symbols are used in exactly the same way, and a combination is used simply for clarification. For instance, $5-\left[3+(2-1)+4\right]$ could be written using only parentheses, but $5-\left(3+(2-1)+4\right)$ would not be as clear at first glance. To avoid any confusion, we alternate the symbols.

RULE

When simplifying an expression having grouping symbols within grouping symbols, perform the operations within the *innermost* set of symbols first.

Example 5   Simplify: $4\left[(3-9)\div(2+1)\right]-3$

Solution   We first remove the parentheses.
$$4\left[-6\div3\right]-3$$
Simplify inside the brackets by dividing.
$$4\left[-2\right]-3$$
Now multiply and subtract.
$$-8-3$$
$$-11 \blacksquare$$

Example 6   Simplify: $5x-2\left[3x-(2x+3)\right]$

Solution   We first remove the parentheses.
$$5x-2\left[3x-2x-3\right] \quad \text{(Notice the sign changes.)}$$
We can now simplify inside the brackets by combining like terms.
$$5x-2\left[x-3\right]$$
We next remove the brackets by multiplying by –2.
$$5x-2x+6$$
Finally combining like terms, we have
$$3x+6. \blacksquare$$

Example 7   Simplify: $2x + 3\left[-4x - (3x - 6)\right]$

Solution $2x + 3\left[-4x - 3x + 6\right]$

$2x + 3\left[-7x + 6\right]$

$2x - 21x + 18$

$-19x + 18$ ∎

EXERCISE 1-7-2

Evaluate.

1. $\left[2 - (6 + 4)\right] \cdot 8 \div 4$

2. $\left[2^3 - 3(4 + 6)\right] \div 2$

3. $5 \cdot 3 - \left[(6 + 2)^2 + 4\right]$

4. $7 - \left[3 + (5 - 1)^2 - 2\right]$

Simplify.

5. $2x + \left[3x - (2x + 1)\right]$

6. $5a + \left[4a - (3a - 2)\right]$

7. $4x - \left[3x - (x + 1)\right]$

8. $10x - \left[x - (3x - 5)\right]$

9. $2 - 3\left[5x - (3x + 1)\right]$

10. $5 - 2\left[4a - (4 - 2a)\right]$

11. $2x - 3\left[1 - (3x + 4)\right]$

12. $5x - 2\left[6 - (4x - 3)\right]$

13. $5-\left[x+(8-3x)\right]$

14. $11-\left[2a+(3a-9)\right]$

15. $\left[3a-(4-2a)\right]-3a$

16. $\left[4x-(3-2x)\right]-x$

17. $8x+2\left[(3x-4)-(x+5)\right]$

18. $4a+3\left[(3a-5)-(2a+1)\right]$

19. $5a-\left[-4-3(2a-3)\right]$

20. $3x-\left[7-2(4x-3)\right]$

21. $2x-3+\left[4x-(x-7)\right]$

22. $5a-2+\left[3a-(a+1)\right]$

23. $7a+4-4\left[3a-(5-2a)\right]$

24. $x-5-2\left[9x-(7-x)\right]$

25. $3(2a+b)+\left[a+3(3a-4b)\right]$

26. $4(x-y)+\left[2x-5(x+y)\right]$

1–8 EVALUATING ALGEBRAIC EXPRESSIONS AND USING FORMULAS

> **OBJECTIVES**
>
> Upon completion of this section you should be able to:
> 1. Evaluate algebraic expressions by substituting numbers for letters.
> 2. Apply this technique in working with formulas.

In an algebraic expression such as $3x$, the variable x is holding the place for some number. As we replace x with various numbers, we will obtain specific values for the expression. For instance, if we replace x with the number 5, we will have

$$3x = 3(5)$$
$$= 15.$$

If we let $x = 8$, then

$$3x = 3(8)$$
$$= 24.$$

If $x = -2$, then

$$3x = 3(-2)$$
$$= -6.$$

The numerical value of an algebraic expression such as $2ab + c$ can be found if we know the values of a, b, and c.

Example 1   Evaluate $2ab + c$ when $a = 3$, $b = -2$, and $c = 15$.

Solution

$$2ab + c = 2(3)(-2) + (15)$$

value of c
value of b
value of a

$$= -12 + 15$$
$$= 3 \quad \blacksquare$$

In an expression such as $x^2 - 2xy + 3x$, the letter x must represent the same number every time it occurs.

A good practice that may help avoid errors is to replace the variables with parentheses and then enter the proper number values within the parentheses.

Example 2   Evaluate $x^2 - 2xy + 3x$ when $x = 2$ and $y = 3$.

Solution
$$
\begin{aligned}
x^2 - 2xy + 3x &= (2)^2 - 2(2)(3) + 3(2) \\
&= 4 - 12 + 6 \\
&= -2 \quad \blacksquare
\end{aligned}
$$

Example 3   Evaluate $x^2 - 2xy + 3x$ when $x = 4$ and $y = -1$.

Solution
$$
\begin{aligned}
x^2 - 2xy + 3x &= (4)^2 - 2(4)(-1) + 3(4) \\
&= 16 + 8 + 12 \\
&= 36 \quad \blacksquare
\end{aligned}
$$

When evaluating an algebraic expression, we must be careful to remember the order of operations that we discussed in the previous section. For instance, the expressions $2x^2$ and $(2x)^2$ would have different values.

Example 4   Evaluate $2x^2$ when $x = -4$.

Solution
$$
\begin{aligned}
2x^2 &= 2(-4)^2 \\
&= 2(16) \\
&= 32 \quad \blacksquare
\end{aligned}
$$

Example 5   Evaluate $(2x)^2$ when $x = -4$.

Solution
$$
\begin{aligned}
(2x)^2 &= (2 \cdot (-4))^2 \\
&= (-8)^2 \\
&= 64 \quad \blacksquare
\end{aligned}
$$

CAUTION   Be careful with an expression such as $-x^2$. Realize that only the value of x is being squared and that the negative sign will precede the result. Recognize that $-x^2$ means $(-1)x^2$.

For example, if $x = 4$, then
$$
\begin{aligned}
-x^2 &= -(4)^2 \\
&= -16.
\end{aligned}
$$

One of the most common uses of evaluating expressions is in working with **formulas**.

Example 6   The perimeter of (distance around) a rectangle can be found by using the formula $P = 2l + 2w$, where P represents the perimeter, l represents the length of the rectangle, and w represents its width.

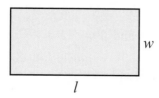

Find the perimeter of a rectangle if the length is 5.4 inches and the width is 3.6 inches.

Solution   We first write the formula for the perimeter.
$$P = 2l + 2w$$
We next make the substitutions $l = 5.4$ in. and $w = 3.6$ in.

$$P = 2(5.4 \text{ in.}) + 2(3.6 \text{ in.})$$
$$P = 10.8 \text{ in.} + 7.2 \text{ in.}$$
$$P = 18.0 \text{ in.} \quad \blacksquare$$

Notice that units are treated like variables.

Example 7   The formula for finding the area A of a trapezoid is $A = \dfrac{1}{2}h(b+c)$, where h represents the height and b and c represent the two bases.

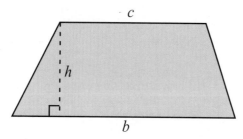

Evaluate $A = \dfrac{1}{2}h(b+c)$ when $h = 6$ meters, $b = 4$ meters, and $c = 1$ meter.

Solution $$A = \frac{1}{2}(6 \text{ m})(4 \text{ m} + 1 \text{ m})$$

$$A = \frac{1}{2}(6 \text{ m})(5 \text{ m})$$

$$A = 15 \text{ m}^2 \quad \blacksquare$$

Some formulas from geometry contain the irrational number π. The number π is a constant that is approximately equal to $\frac{22}{7}$. We write $\pi \approx \frac{22}{7}$.

Example 8   The volume V of a right circular cone where the height is h and the radius of the base is r is found by the formula

$$V = \frac{1}{3}\pi r^2 h.$$

Approximate the volume if $r = 3$ cm and $h = 14$ cm.

Solution

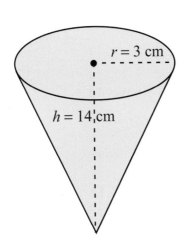

$$V \approx \frac{1}{3}\left(\frac{22}{7}\right)(3\text{ cm})^2(14\text{ cm})$$

$$V \approx \frac{1}{3}\left(\frac{22}{7}\right)(9\text{ cm}^2)(14\text{ cm})$$

$$V \approx 132\text{ cm}^3 \quad\blacksquare$$

Example 9   A baseball is thrown upward with an initial velocity v of $40\,\dfrac{\text{ft}}{\text{sec}}$. The distance s of the ball above the ground at t seconds is given by $s = vt + \dfrac{1}{2}gt^2$, where g is the acceleration due to gravity. Find the distance of the ball above the ground at the end of 2 seconds. Use $g = -32\,\dfrac{\text{ft}}{\text{sec}^2}$.

Solution   Using the formula $s = vt + \dfrac{1}{2}gt^2$ and making the substitutions $v = 40\,\dfrac{\text{ft}}{\text{sec}}$, $t = 2$ sec, and $g = -32\,\dfrac{\text{ft}}{\text{sec}^2}$, we obtain

$$s = vt + \frac{1}{2}gt^2$$

$$s = \left(40\,\frac{\text{ft}}{\text{sec}}\right)(2\text{ sec}) + \frac{1}{2}\left(-32\,\frac{\text{ft}}{\text{sec}^2}\right)(2\text{ sec})^2$$

$$s = \left(40\,\frac{\text{ft}}{\text{sec}}\right)(2\text{ sec}) + \frac{1}{2}\left(-32\,\frac{\text{ft}}{\text{sec}^2}\right)(4\text{ sec}^2)$$

$$s = 80\text{ ft} - 64\text{ ft}$$

$$s = 16\text{ ft} \quad\blacksquare$$

EXERCISE 1-8-1

Evaluate.

1. $3x + 5$ when $x = -7$

2. $16 - 2x$ when $x = -3$

3. x^2 when $x = -5$

4. $-x^2$ when $x = -5$

5. $(-x)^2$ when $x = -5$

6. $5x^2$ when $x = 4$

7. $(5x)^2$ when $x = 4$

8. $-x^2 + 3x - 1$ when $x = -2$

9. $3x^2 - 5x + 2$ when $x = 3$

10. $2x^2 - 3x + 1$ when $x = -3$

11. $-x^2 + xy - 6$ when $x = -5$ and $y = 1$

12. $2x^3 + 3xyz - 2z^2$ when $x = 2$, $y = 5$ and $z = -1$

13. The perimeter P of a rectangle is given by $P = 2l + 2w$, where l and w represent the length and width. Find P when $l = 12$ cm and $w = 7$ cm.

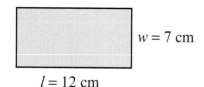

$w = 7$ cm

$l = 12$ cm

14. The area A of a rectangle is given by $A = lw$, where l represents the length and w represents the width. Find A when $l = 9$ ft and $w = 3$ ft.

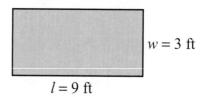

$w = 3$ ft

$l = 9$ ft

15. The perimeter P of a square is given by $P = 4s$, where s represents the length of one side. Find P when $s = 8$ meters.

$s = 8$ m

16. The area A of a square is given by $A = s^2$. Find A when $s = 4$ inches.

$s = 4$ in

17. The perimeter P of a triangle is given by the formula $P = a + b + c$, where a, b, and c represent the three sides of the triangle. Find P when $a = 14.5$ ft, $b = 10.6$ ft, and $c = 17.8$ ft

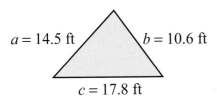

$a = 14.5$ ft $b = 10.6$ ft

$c = 17.8$ ft

18. The area A of a triangle is given by $A = \dfrac{1}{2}bh$, where b represents the base and h represents the height. Find A if $b = 7$ in. and $h = 6$ in.

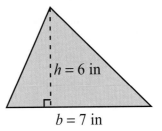

$h = 6$ in

$b = 7$ in

19. A distance formula from physics is $d = rt$, where r represents the rate and t represents the time. Find d when $r = 55 \dfrac{\text{mi}}{\text{hr}}$ and $t = 4$ hours.

20. A formula from business for finding interest is $I = Prt$, where P represents the principal amount invested, r represents the rate of interest, and t represents the time invested. Find I when $P = \$8000$, $r = 8\%$ per year, and $t = 3$ years.

21. The circumference C of a circle is given by $C = \pi d$, where d represents the diameter of the circle. Approximate C for $\pi \approx \dfrac{22}{7}$ and $d = 14$ meters.

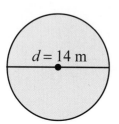

22. The area A of a circle is given by $A = \pi r^2$, where r represents the radius of the circle. Approximate A for $\pi \approx \dfrac{22}{7}$ and $r = 7$ inches.

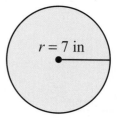

23. A formula for changing Fahrenheit temperature to Celsius is given by $C = \dfrac{5}{9}(F - 32°)$. Find C when $F = 68°$.

24. A formula for changing Celsius temperature to Fahrenheit is $F = \dfrac{9}{5}C + 32°$. Find F when $C = 30°$.

25. The volume of a rectangular solid is given by $V = lwh$. Find V when $l = 3$ cm, $w = 1.5$ cm, and $h = 4$ cm.

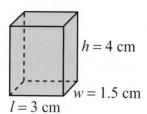

26. The volume of a cube is given by $V = s^3$. Find V when $s = 5$ meters.

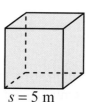

27. The volume of a right circular cylinder of height h having a circular base with radius r is given by $V = \pi r^2 h$. Find V for $\pi \approx \dfrac{22}{7}$, $r = 7$ ft, and $h = 3$ ft.

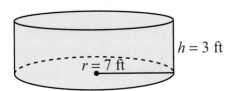

28. The area of a trapezoid is given by $A = \dfrac{1}{2} h(b+c)$. Find A when $h = 9$ in., $b = 13$ in., and $c = 7$ in.

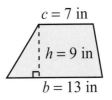

29. The volume of a right circular cone is given by $V = \dfrac{1}{3} \pi r^2 h$. Find V for $\pi \approx \dfrac{22}{7}$, $r = 7$ meters, and $h = 9$ meters.

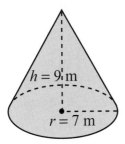

30. The volume of a sphere is given by $V = \dfrac{4}{3} \pi r^3$. Find V for $\pi \approx \dfrac{22}{7}$ and $r = 21$ meters.

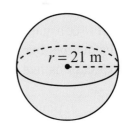

31. If an object is thrown downward from a height h with an initial velocity v, then the distance s of the object above the ground at any time t can be found using the formula $s = h + vt + \dfrac{1}{2} gt^2$, where g is the acceleration due to gravity. Find s when $h = 5500$ cm, $v = -100 \dfrac{\text{cm}}{\text{sec}}$, $t = 2$ seconds, and $g = -980 \dfrac{\text{cm}}{\text{sec}^2}$.

32. A formula for determining the total amount A in an account when a principal P is invested at a rate of interest r for a time t is $A = P(1 + rt)$. Find the total amount in the account at the end of 5 years if $10,000 was invested at 8% annual interest.

CHAPTER 1 SUMMARY

The number in brackets refers to the section of the chapter that discusses the concept.

Terminology

- Elementary algebra is a study of the **real numbers** and their properties. [1-1]
- Positive and negative real numbers are referred to as **signed numbers**. [1-1]
- The **negative** (**opposite** or **additive inverse**) of a given number is a number that, when added to the given number, yields a sum of zero. [1-1]
- The **real number line** is used to show a correspondence between the real numbers and the points of the line. [1-1]
- The **absolute value** of a number is its distance from zero on the number line. [1-1]
- **Factors** are expressions that are multiplied. [1-4]
- A whole number **exponent** indicates the number of times a **base** is used as a factor. [1-4]
- A numerical factor is referred to as the **coefficient**. [1-4]
- **Variables** are letters that are used to represent numbers. [1-5]
- An **algebraic expression** is an expression in which letters are used to represent numbers. [1-5]
- In an indicated sum or difference, the expressions being added or subtracted are called **terms**. [1-5]
- **Similar terms** have the same arrangement of variables and exponents. [1-5]
- A **polynomial** is an algebraic expression having one or more terms that contain only whole number exponents. [1-6]
- A **monomial** is a polynomial having one term. [1-6]
- A **binomial** is a polynomial having two terms. [1-6]
- A **trinomial** is a polynomial having three terms. [1-6]
- Parentheses, brackets, and braces are used as **grouping symbols**. [1-7]

Rules and Procedures

Signed Numbers

- To add signed numbers having like signs, add the absolute values of the numbers and use the common sign. [1-1]
- To add signed numbers with unlike signs, subtract the absolute values of the numbers. The sign of the number with the larger absolute value is the sign of the answer. [1-1]
- To subtract one signed number from another, change the sign of the number being subtracted and use the appropriate rule for addition. [1-2]
- The product or quotient of two signed numbers having like signs will be positive. [1-3]
- The product or quotient of two signed numbers having unlike signs will be negative. [1-3]

Laws of Exponents

- $x^a \cdot x^b = x^{a+b}$ [1-4]
- For $x \neq 0$,

  $\dfrac{x^a}{x^b} = x^{a-b}$ if a is *greater* than b,

  $\dfrac{x^a}{x^b} = \dfrac{1}{x^{b-a}}$ if a is *less* than b, and

  $\dfrac{x^a}{x^b} = 1$ if $a = b$. [1-4]

Similar Terms

- Only similar terms can be added or subtracted. [1-5]
- To add or subtract similar terms, combine the coefficients and use this result as the coefficient of the common variable factors. [1-5]

Multiplying Polynomials by Monomials

- To multiply a polynomial by a monomial, multiply each term of the polynomial by the monomial. [1-6]

Order of Operations

- If an expression contains only additions and subtractions, these operations are performed in order from left to right. [1-2]

- If no grouping symbols occur in an expression to be evaluated, exponents must be worked first, then multiplication and division (left to right), and then addition and subtraction (left to right). [1-7]
- When simplifying an expression having grouping symbols within grouping symbols, remove the innermost set of grouping symbols first. [1-7]

Evaluating Algebraic Expressions

- To evaluate an algebraic expression, substitute the correct number for each letter and then evaluate the resulting numerical expression. [1-8]

CHAPTER 1 REVIEW

Classify each of the following numbers as counting, whole, integer, rational, or irrational. A number may have more than one classification.

1. -6

2. $\dfrac{3}{4}$

3. $\dfrac{15}{3}$

Evaluate.

4. $\left|-\sqrt{3}\right|$

5. $7+(-9)$

6. $(-17)+(-5)$

7. $\dfrac{5}{9}+\left(-\dfrac{3}{9}\right)$

8. $-5+8+(-10)$

9. $6+(-15)+4$

10. $5-8+2$

11. $4-(-6)-1$

12. $-3+(-4)+9$

13. $-13+5-(-16)-2$

14. $-4+9+2-8-7+6$

15. $(8)(-5)(-7)$

16. $(-4)(3)(10)$

17. $\left(\dfrac{4}{9}\right)\div\left(-\dfrac{2}{3}\right)$

18. $\dfrac{-32}{-8}$

19. $(-3)(-1)(5)(-10)\left(-\dfrac{2}{5}\right)(3)$

20. $2(3)+6(4)$

21. $3\cdot9\div3\cdot2$

Simplify.

22. $\left(x^3\right)\left(x^7\right)$

23. $(-2x)\left(4x^3\right)(x)\left(-3x^2\right)$

24. $\left(3x^2y\right)\left(5xy^3z\right)\left(-2y^2z\right)$

25. $\dfrac{9x^7y^3}{-3x^2y}$

26. $\dfrac{8x^4y^4}{2xy^6}$

27. $9a - 3a$

28. $5x - 8x - x$

29. $3x^2 + 2x - 2x^2 + x^2$

30. $6xy^2 + 7x^2y - 3xy^2$

31. $-13ab + 5a - 7ab + 2$

Multiply.

32. $5(2x - 4y)$

33. $-\left(3x^2 + 2x - 5\right)$

34. $6x\left(2x^2 - 3x - 4\right)$

35. $3xy\left(2x^2y - 5xy^3 + 2\right)$

36. $-2abc\left(3a^2b - 5abc^2 - bc\right)$

Simplify.

37. $5x^2 + 3(2x-1) - 4x(2x^2 - 3x - 1)$

38. $2x[3x - 2(x+3) + x]$

39. $3x^2 - 2x\{2x - 4[3 - 4(x-5)]\}$

Evaluate.

40. $3x^2$ when $x = -5$

41. $-x^3$ when $x = -2$

42. $2x^2 - 3x + 1$ when $x = -4$

43. $5x^3 - 2xy^2 - 3y$ when $x = -1$ and $y = 2$

44. Using the formula $V = \dfrac{1}{3}\pi r^2 h$, approximate V when $\pi \approx \dfrac{22}{7}$, $r = 2$ cm, and $h = 21$ cm.

CHAPTER 1 PRACTICE TEST

1. Give the negative or opposite of $-\dfrac{3}{5}$.

1. _____

Evaluate.

2. $-|-8|$

2. _____

3. $(-24)+(-16)$

3. _____

4. $\left(-\dfrac{7}{11}\right)+\left(\dfrac{5}{11}\right)$

4. _____

5. $18-(-23)$

5. _____

6. $15+(-5)-(-31)$

6. _____

7. $-3-11+21+9-25-7$

7. _____

8. $\left(\dfrac{2}{3}\right)(-42)$

8. _____

9. $\left(-\dfrac{1}{3}\right) \div \left(-\dfrac{5}{6}\right)$

9. _____

10. $-11\left(\dfrac{1}{3}\right)(-42)\left(-\dfrac{2}{7}\right)$

10. _____

11. $(-3)^5$

11. _____

Simplify.

12. $\left(3x^2y^3\right)\left(-2xy^2\right)\left(4x^3y^2\right)$

12. _____

13. $\dfrac{33a^6}{99a^4}$

13. _____

14. $\dfrac{12x^5y^3}{-3xy^4}$

14. _____

15. $13x - 5y - 8y$

15. _____

16. $3a^2b - 4ab^2 + 6a^2b + 10ab^2 - 8a^2b + 7ab$

16. _____

17. $25x - (4x + 3y - 2)$ **17.** _____

18. $3x^2 - \left[5x - 4x(x - 3) \right]$ **18.** _____

19. Multiply: $4a^2b^2 \left(5a - 2a^2b - 5b^2 \right)$ **19.** _____

20. If $x = -2$, evaluate $-x^2 - 15x + 4$.   **20.** _____

21. Using the formula $s = h + \dfrac{1}{2}gt^2$, find

s when $h = 1996$ cm, $g = -980\ \dfrac{\text{cm}}{\text{sec}}$,

and $t = 2$ sec.   **21.** _____

CHAPTER 2

SURVEY

The following questions refer to material discussed in this chapter. Work as many problems as you can and check your answers with the answer key in the back of the book. The results will direct you to the sections of the chapter in which you need to work. If you answer all questions correctly, you may already have a good understanding of the material contained in this chapter.

1. Are $x + 4 = 2x + 1$ and $x = 3$ equivalent equations?

1. _____

Solve for x.

2. $x + 5 = 3$

2. _____

3. $-3x = 12$

3. _____

4. $\frac{1}{2}x = 6$

4. _____

5. $\frac{x}{3} = \frac{3}{7}$

5. _____

6. $x - 7 = 4x + 5$

6. _____

7. $2x - 3(x - 2) = \dfrac{1}{2}(x + 1)$

7. _____

8. $3(2a - 3x) = 4(a + 2x)$

8. _____

9. $\dfrac{2}{5}x + 2 \geq \dfrac{x}{2} + 1$

9. _____

10. $|x - 3| > 5$

10. _____

First-Degree Equations and Inequalities: One Unknown

The manipulative skills of algebra ultimately lead to the solution of equations. Equations give answers to problems, and the solving of problems is the central theme of algebra.

In this chapter we will study the techniques of solving first-degree equations in one unknown. The fundamental concepts that you learned in the previous chapter supply the necessary tools for this chapter. Other types of equations will be studied once we have mastered further techniques necessary for solving them.

2–1   CONDITIONAL EQUATIONS

> **OBJECTIVES**
>
> Upon completion of this section you should be able to:
> 1. Identify equations as true, false, or conditional.
> 2. Identify equivalent equations.

An **equation** in mathematics states that two numerical expressions are equal. The following three statements are all equations. The first is false, the second is true, and the third is neither true nor false.

1. $5 + 2 = 9$ *false*
2. $5 + 4 = 9$ *true*
3. $5 + x = 9$ *neither true nor false*

Equations such as number 3 above are called conditional equations. **Conditional equations** contain variables. In this chapter we will work with conditional equations.

In an equation such as $5 + x = 9$, the letter x holds the place for a numerical value. Since x can take on various values, we refer to this letter as a **variable**. The variable can also be referred to as the **unknown**. Depending on the numerical value we substitute for the variable, we will obtain either a true statement or a false statement. To **solve** a conditional equation we must find a replacement for the variable that will make the equation a true statement. Such a replacement is called a **solution** to the equation.

Example 1   Determine if 4 is a solution for $5 + x = 9$.

Solution   Replace x with 4 and simplify.

$$5 + (4) = 9$$
$$9 = 9$$

Since the resulting equation is true, 4 is a solution. ■

EXERCISE 2-1-1

Identify the following equations as true, false, or conditional.

1. $5 + 6 = 12 - 1$ **2.** $2 + 7 = 8 - 1$ **3.** $x - 3 = 5$

4. $27 - 21 = 17 - 23$ **5.** $x = 8$

Replace the variable x with the number 3 and determine if 3 is a solution.

6. $x + 17 = 20$ **7.** $x - 8 = -11 + 6$ **8.** $x + 4 = 2x - 1$

9. $2x - 9 = -6 + x$ **10.** $x - 6 = 2x - 3$

Two or more equations are **equivalent equations** if their solutions are identical. For instance, $2x + 1 = 5$ and $3x + 1 = 7$ both have a solution of 2 and are therefore equivalent equations. The equations $5x - 1 = 14$ and $x = 3$ both have a solution of 3 and are therefore equivalent.

Example 2   Determine if $3x - 3 = x + 7$ and $x = 5$ are equivalent equations.

Solution   The second equation states $x = 5$. Substitute 5 for x in the first equation.
$$3(5) - 3 = 5 + 7$$
$$12 = 12$$

Since this last equation is a true statement, 5 is a solution to both equations and we conclude that the equations are equivalent. ∎

Example 3   Determine if $-y + 6 = 2y - 1$ and $y = -2$ are equivalent.

Solution   Substitute -2 for y in the first equation.
$$-(-2) + 6 = 2(-2) - 1$$
$$8 = -5$$

Since this last equation is false, -2 is not a solution to both equations and we conclude that these equations are not equivalent. ∎

EXERCISE 2-1-2

Determine if each of the following pairs of equations are equivalent.

1. $x + 7 = 2x + 6$ and $x = 1$ 2. $3x - 5 = 2x + 1$ and $x = 5$

3. $2a + 1 = a$ and $a = 1$ 4. $4a - 3 = a + 6$ and $a = 3$

5. $5x - 4 = 3x + 2$ and $x = -3$ **6.** $3 - 2x = 5 - x$ and $x = -2$

7. $2n - 2 + n = -1 + n$ and $n = \dfrac{1}{3}$ **8.** $\dfrac{3}{4}n + 1 = n - \dfrac{5}{4} + 4n$ and $n = 1$

9. $3x + 1 = x + 5$ and $x = 2$ **10.** $2x + 17 = 10 - 5x$ and $x = -1$

2–2   RULES FOR SOLVING FIRST-DEGREE EQUATIONS

> **OBJECTIVE**
>
> Upon completion of this section you should be able to solve simple first-degree equations.

The **degree** of a polynomial equation in one unknown is the highest power (largest exponent) of the unknown in any term. For instance, $x^2 = 9$ is a *second-degree equation* in x since 2 is the highest power of x, and $x^3 + 2x^2 - x = 14$ is a *third-degree equation* in x since 3 is the highest power of x. Equations in which the highest power of the unknown is 1 are **first-degree equations**, which we will study in this chapter.

An example of the simplest type of first-degree equation is $x = 5$. Clearly, this equation is conditional since it is true only if x is replaced by 5.

The methods for solving equations involve the changing of a more complicated equation, such as $2x - 4 = x + 1$, to a simpler equivalent equation, such as $x = 5$. The question now becomes, "How may an equation be changed without changing the value of its solution?"

To answer this question, we need to consider how the arithmetic operations affect the equality of two expressions. If two numbers are equal, such as $4 = 4$, we could add the same number, such as 3, to each side and the resulting numbers would still be equal.

$$4 + 3 = 4 + 3$$
$$7 = 7$$

We could subtract the same number from each side and the resulting numbers would be equal.

$$4 - 3 = 4 - 3$$
$$1 = 1$$

We could multiply each side by the same number and the results would be equal.

$$4(5) = 4(5)$$
$$20 = 20$$

Finally, we could divide each side by the same nonzero number and the results would be equal.

$$4 \div 2 = 4 \div 2$$
$$2 = 2$$

In each case the resulting quantities are equal.

An equation is a statement that two quantities are equal. So we can use the arithmetic operations to rewrite an equation to change it to a simpler equation. First, we will state a rule for addition and subtraction.

> ### RULE
>
> If any number is added to or subtracted from both sides of an equation, the resulting equation will be equivalent to the original equation.

Example 1 $3x + 5 = 2x - 3$ is equivalent to $3x + 5 + 10 = 2x - 3 + 10$. (10 has been added to both sides of the equation.) ∎

Of course, our object is to change an equation like the one above into an equation where the variable is on one side and a number is on the other side. We therefore carefully choose the number we wish to add to or subtract from both sides.

Example 2   Solve for x: $x - 7 = 3$

Solution   Since we wish to arrive at an equation having only the unknown x on one side, we will add 7 to both sides.

$$x - 7 + 7 = 3 + 7$$
$$x = 10 \quad ∎$$

It is important to check to see if your solution is correct. Do this by substituting your value for the variable into the original equation to see if the equation is now true. For Example 2, the check would be as follows.

$$\textit{Check} \qquad (10) - 7 = 3$$
$$3 = 3$$

Example 3   Solve for x: $3x + 5 = 2x - 3$

Solution   Subtract 5 from both sides.

$$3x + 5 - 5 = 2x - 3 - 5$$
$$3x = 2x - 8$$

Subtract $2x$ from both sides.

$$3x - 2x = 2x - 8 - 2x$$
$$x = -8$$

$$\textit{Check} \qquad 3(-8) + 5 = 2(-8) - 3$$
$$-24 + 5 = -16 - 3$$
$$-19 = -19$$

Since substituting -8 for x results in a true equation, -8 is the correct solution. ∎

EXERCISE 2-2-1

Solve for x. Check.

1. $x + 3 = 5$

2. $x + 7 = 2$

3. $x - 4 = 6$

4. $x - 8 = -11$

5. $x - 1 = 1$

6. $x + 5 = 5$

7. $x + 7 = -7$

8. $3x + 2 = 2x + 7$

9. $2x - 9 = 4 + x$

10. $5x - 2 = 4x - 5$

We will now state a rule for multiplication and division.

> **RULE**
>
> If both sides of an equation are multiplied or divided by the same nonzero number, the resulting equation is equivalent to the original equation.

Note that the rule states that when we use division to solve an equation, we must divide by a nonzero number. This condition exists since division by zero is undefined. The rule also states that when we use multiplication to solve an equation, we must multiply by a nonzero number. Multiplying both sides of an equation by zero always yields $0 = 0$, which is true but not helpful in solving for the variable.

Example 4   Solve for x: $2x = 10$

Solution   Divide each side by 2.

$$\frac{2x}{2} = \frac{10}{2}$$

$$x = 5$$

Check $2(5) = 10$

$10 = 10$ ■

Example 5   Solve for x: $-3x = 15$

Solution   Divide each side by -3.

$$\frac{-3x}{-3} = \frac{15}{-3}$$

$$x = -5$$

Check $-3(-5) = 15$

$15 = 15$ ■

Example 6   Solve for x: $\frac{1}{3}x = 5$

Solution   Multiply each side by 3 $\left(\text{the reciprocal of } \frac{1}{3}\right)$.

$$3\left(\frac{1}{3}x\right) = 3(5)$$

$$x = 15$$

Check $\frac{1}{3}(15) = 5$

$5 = 5$ ■

Example 7   Solve for x: $\frac{x}{4} = \frac{5}{12}$

Solution   Multiply both sides by 4.

$$4\left(\frac{x}{4}\right) = 4\left(\frac{5}{12}\right)$$

$$x = \frac{5}{3}$$

Check

$$\frac{\frac{5}{3}}{4} = \frac{5}{12}$$

$$\frac{5}{3} \cdot \frac{1}{4} = \frac{5}{12}$$

$$\frac{5}{12} = \frac{5}{12} \quad ■$$

EXERCISE 2-2-2

Solve for x. Check your answers.

1. $2x = 24$ 2. $3x = 12$ 3. $5x = 2$

4. $3x = 1$

5. $4x = -2$

6. $7x = \dfrac{7}{2}$

7. $\dfrac{3}{4} = -9x$

8. $6 = -2x$

9. $-x = -5$

10. $-12x = -4$

11. $\dfrac{1}{4}x = 3$

12. $\dfrac{1}{5}x = 2$

13. $\dfrac{x}{6} = -1$

14. $\dfrac{x}{3} = \dfrac{-3}{7}$

15. $-25 = \dfrac{1}{5}x$

16. $-3 = \dfrac{1}{9}x$

17. $-\dfrac{x}{4} = -16$

18. $-\dfrac{1}{2}x = -\dfrac{3}{4}$

19. $\dfrac{2x}{3} = \dfrac{4}{5}$

20. $\dfrac{3x}{4} = \dfrac{3}{2}$

2–3   COMBINING RULES TO SOLVE FIRST-DEGREE EQUATIONS

> **OBJECTIVES**
>
> Upon completion of this section you should be able to:
> 1. Solve equations that require more than one operation.
> 2. Apply a step-by-step procedure to solve first-degree equations.

Most equations you will encounter will involve some combination of the operations of addition, subtraction, multiplication, and division. They may also involve parentheses, combining like terms, and many of the other topics developed in Chapter 1.

Example 1   Solve for x: $12 + 3(x - 1) = x + 3$

Solution   We observe that it will take more than one operation to find the solution. Where do we begin? We begin by using the order of operations to simplify the expression on the left side of the equation.

$$12 + 3(x - 1) = x + 3$$
$$12 + 3x - 3 = x + 3$$
$$3x + 9 = x + 3$$

Now we apply the rules for changing to an equivalent equation.
We subtract x from each side.

$$3x + 9 - x = x + 3 - x$$
$$2x + 9 = 3$$

Now we subtract 9 from each side.

$$2x + 9 - 9 = 3 - 9$$
$$2x = -6$$

Finally, we divide both sides by 2.

$$\frac{2x}{2} = \frac{-6}{2}$$

So the result is $x = -3$.

Check
$$12 + 3((-3) - 1) = (-3) + 3$$
$$12 + 3(-4) = 0$$
$$12 + (-12) = 0$$
$$0 = 0 \quad \blacksquare$$

You will find that the following step-by-step procedure for solving equations will help you avoid mistakes as you find a solution.

A PROCEDURE FOR SOLVING FIRST-DEGREE EQUATIONS

1. Clear grouping symbols using the appropriate operation.
2. Multiply both sides by the least common denominator of all fractions appearing in the equation.
3. Combine similar terms on each side of the equation.
4. Add or subtract terms on both sides of the equation to get the unknown on one side and everything else on the other.
5. Divide both sides of the equation by the coefficient of the unknown.
6. Simplify the solution.

Example 2   Solve for x: $\dfrac{x}{3} - 2 = \dfrac{1}{2}(1 - x)$

Solution   The first step is to clear the parentheses using the distributive property.

$$\frac{x}{3} - 2 = \frac{1}{2} - \frac{x}{2}$$

Next we clear fractions by multiplying each side of the equation by the LCD (least common denominator), which is 6.

$$6\left(\frac{x}{3} - 2\right) = 6\left(\frac{1}{2} - \frac{x}{2}\right)$$

$$6\left(\frac{x}{3}\right) - 6(2) = 6\left(\frac{1}{2}\right) - 6\left(\frac{x}{2}\right)$$

$$2x - 12 = 3 - 3x$$

Next we add $3x$ to each side.

$$2x - 12 + 3x = 3 - 3x + 3x$$

$$5x - 12 = 3$$

Then we add 12 to both sides.

$$5x - 12 + 12 = 3 + 12$$

$$5x = 15$$

Finally, we divide each side by 5.

$$\frac{5x}{5} = \frac{15}{5}$$

$$x = 3$$

Check $\dfrac{(3)}{3} - 2 = \dfrac{1}{2}(1 - (3))$

$1 - 2 = \dfrac{1}{2}(-2)$

$-1 = -1$ ∎

CAUTION   In an equation involving fractions, make sure you multiply *every* term by the least common denominator, not just the terms containing fractions. In Example 2, we multiplied every term by 6 to clear the fractions.

Example 3   Solve for x: $\dfrac{5}{7}x - x - \dfrac{5}{3} = \dfrac{1}{21}(3x - 38)$

Solution   Clear parentheses.

$$\frac{5}{7}x - x - \frac{5}{3} = \frac{3x}{21} - \frac{38}{21}$$

Multiply both sides by 21.

$$21\left(\frac{5}{7}x - x - \frac{5}{3}\right) = 21\left(\frac{3x}{21} - \frac{38}{21}\right)$$

$$21\left(\frac{5}{7}x\right) - 21(x) - 21\left(\frac{5}{3}\right) = 21\left(\frac{3x}{21}\right) - 21\left(\frac{38}{21}\right)$$

$$15x - 21x - 35 = 3x - 38$$

Combine similar terms.

$$-6x - 35 = 3x - 38$$

Add $6x$ to both sides.

$$-6x - 35 + 6x = 3x - 38 + 6x$$
$$-35 = -38 + 9x$$

Add 38 to both sides.

$$-35 + 38 = -38 + 9x + 38$$
$$3 = 9x$$

Divide both sides by 9.

$$\frac{3}{9} = \frac{9x}{9}$$

$$\frac{1}{3} = x \quad \text{This could also be written as } x = \frac{1}{3}.$$

Check $$\frac{5}{7}\left(\frac{1}{3}\right) - \left(\frac{1}{3}\right) - \frac{5}{3} = \frac{1}{21}\left(3\left(\frac{1}{3}\right) - 38\right)$$

$$\frac{5}{21} - \frac{1}{3} - \frac{5}{3} = \frac{1}{21}(1 - 38)$$

$$\frac{5}{21} - \frac{6}{3} = \frac{1}{21}(-37)$$

$$\frac{5}{21} - \frac{42}{21} = -\frac{37}{21}$$

$$-\frac{37}{21} = -\frac{37}{21} \quad \blacksquare$$

Example 4   Solve for a: $\dfrac{2}{3} - \dfrac{a-2}{6} = \dfrac{1}{2}$

Solution   To clear the fractions, multiply both sides by the least common denominator, 6.

$$6\left(\dfrac{2}{3} - \dfrac{a-2}{6}\right) = 6\left(\dfrac{1}{2}\right)$$

Note that $\dfrac{a-2}{6}$ is one term. The fraction bar is a grouping symbol. When it is removed we place the numerator, $(a-2)$, in parentheses.

$$6\left(\dfrac{2}{3}\right) - 6\left(\dfrac{a-2}{6}\right) = 6\left(\dfrac{1}{2}\right)$$

$$4 - (a-2) = 3$$
$$4 - a + 2 = 3$$
$$6 - a = 3$$
$$-a = -3$$
$$a = 3$$

Check

$$\dfrac{2}{3} - \dfrac{3-2}{6} = \dfrac{1}{2}$$
$$\dfrac{2}{3} - \dfrac{1}{6} = \dfrac{1}{2}$$
$$\dfrac{1}{2} = \dfrac{1}{2}$$ ∎

Example 5   The perimeter P of a rectangle is given by $P = 2l + 2w$, where l and w represent the length and width. Find l when $P = 52$ inches and $w = 6$ inches.

Solution

$$P = 2l + 2w$$
$$52 = 2l + 2(6)$$
$$52 = 2l + 12$$
$$52 - 12 = 2l + 12 - 12$$
$$40 = 2l$$
$$\dfrac{40}{2} = \dfrac{2l}{2}$$
$$20 = l$$

$w = 6$ in
$l = ?$

The length is 20 inches. ∎

EXERCISE 2-3-1

Solve and check.

1. $3x + 5 = x + 7$

2. $4x - 3 = x - 9$

3. $3y - 1 = 2(y - 5)$

4. $3y + 2 = 5y - 8$ **5.** $a - 7 = 4a + 5$ **6.** $2a + 3 = 3a + 5$

7. $x - 6 = 5x - 14$ **8.** $3(x - 4) = 5 + 2(x + 1)$ **9.** $5x - 4 = 2x + 6$

10. $3x + 7 = 5x - 4$ **11.** $3y + 2(y - 5) = 7 - (y + 3)$ **12.** $5y - 3(y + 1) = 5$

13. $7x + 5 - 2(x - 1) = 21$ **14.** $5 + 6x - 3 = 2 + 4x$ **15.** $3(y + 1) + 3 = 7(y - 2)$

16. $5x - (2x - 3) = 4(x + 9)$ **17.** $\dfrac{x}{2} = \dfrac{1}{5} - x$ **18.** $4 - \dfrac{y + 2}{3} = 1$

19. $b - \dfrac{1}{2} = \dfrac{b}{3} + 7$ **20.** $\dfrac{b - 3}{4} - 9 = 11$ **21.** $5(x + 4) = \dfrac{1}{3}(x - 10)$

22. $\dfrac{2x + 5}{3} = 0$ **23.** $\dfrac{2}{3} - 3(x - 1) = \dfrac{3}{5}$ **24.** $2x - 3(x - 2) = \dfrac{1}{2}(x + 1)$

25. $\dfrac{2}{3} + x = 1 - \dfrac{x - 1}{2}$ **26.** $4y - 2(y - 3) = \dfrac{1}{4}(y + 3)$ **27.** $x - \dfrac{1}{5}x + 1 = \dfrac{1}{3}(x - 5)$

28. $\frac{2}{3}y + \frac{y}{2} - \frac{1}{2} = \frac{1}{3}(y - 14)$ **29.** $\frac{2}{3}x - \frac{1}{2}x = x + \frac{1}{6}$ **30.** $\frac{x}{5} - \frac{2}{3}x + \frac{1}{2} = \frac{1}{3}(x - 4)$

31. The perimeter of a rectangle is given by $P = 2l + 2w$. Find w when $P = 84$ meters and $l = 31$ meters.

32. The area of a trapezoid is given by $A = \frac{1}{2}h(b + c)$. Find b when $h = 10$ inches, $c = 3$ inches, and $A = 35$ square inches.

33. The circumference of a circle is given by $C = \pi d$. Approximate the value for d if $\pi \approx \frac{22}{7}$ and $C = 154$ yards.

34. The volume of a cylinder is given by $V = \pi r^2 h$. Approximate the value for h when $r = 7$ cm, $\pi \approx \frac{22}{7}$, and $V = 77$ cm^3.

2–4   LITERAL EQUATIONS

> **OBJECTIVES**
>
> Upon completion of this section you should be able to:
> 1. Identify a literal equation.
> 2. Apply the rules of the previous section to solve literal equations.

An equation having more than one letter is sometimes called a **literal equation**. We occasionally need to solve such an equation for one letter in terms of the others. We will use the same steps for solving an equation as we did in the previous section.

Example 1   Solve for x: $3abx + cy = 2abx + 4cy$

Solution   First subtract $2abx$ from both sides.

$$3abx + cy - 2abx = 2abx + 4cy - 2abx$$

$$abx + cy = 4cy$$

Subtract cy from both sides.

$$abx + cy - cy = 4cy - cy$$

$$abx = 3cy$$

Divide both sides by ab.

$$\frac{abx}{ab} = \frac{3cy}{ab}$$

or $\qquad x = \dfrac{3cy}{ab}$ ∎

Example 2   Solve for y: $\dfrac{2}{3}(x + y) = 3(y + a)$

Solution   First clear parentheses.

$$\frac{2}{3}x + \frac{2}{3}y = 3y + 3a$$

Multiply both sides by 3.

$$3\left(\frac{2}{3}x + \frac{2}{3}y\right) = 3(3y + 3a)$$

$$2x + 2y = 9y + 9a$$

Subtract $9y$ from both sides.

$$2x - 7y = 9a$$

Subtract $2x$ from both sides.

$$-7y = 9a - 2x$$

Divide both sides by -7.

$$y = \frac{9a - 2x}{-7} \quad \blacksquare$$

Sometimes the form of an answer may be changed without changing the value of the answer. In Example 2 we could multiply both numerator and denominator by -1 to change the form of the answer.

$$y = \frac{9a - 2x}{-7} = \frac{(-1)(9a - 2x)}{(-1)(-7)} = \frac{-9a + 2x}{7} = \frac{2x - 9a}{7}$$

The advantage of this last expression over the first is that there are not as many negative signs in the answer.

The most commonly used literal equations are formulas from geometry, physics, business, electronics, and so on.

Example 3   The formula for the area A of a trapezoid is $A = \frac{1}{2}h(b+c)$, where h is the distance between the two parallel sides b and c. Solve for c.

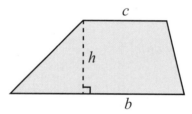

Solution   First remove parentheses.

$$A = \frac{1}{2}hb + \frac{1}{2}hc$$

Multiply both sides by 2.

$$2A = 2\left(\frac{1}{2}hb + \frac{1}{2}hc\right)$$

$$2A = hb + hc$$

Subtract hb from both sides.

$$2A - hb = hc$$

Divide both sides by h.

$$\frac{2A - hb}{h} = c$$

This could also be written as $c = \frac{2A - hb}{h}$. $\blacksquare$

EXERCISE 2-4-1

Solve for x.

1. $2x + y = x + 3y$

2. $3x - 2y = x + 4y$

3. $3x + 2y = 6y + x$

4. $2x + 8y = 5x - y$

5. $2(x + a) = x - 4a$

6. $3x + a = 7(x - a)$

7. $\dfrac{2}{5}(x - a) = 4(x + a)$

8. $4a - 2x = 3(4x - 8a)$

9. $b - 5x = 3(x + b)$

Solve for *y*.

10. $3(2x+y)=19y+4x$

11. $11x+15y=5(3x-y)$

12. $4(x-2y)+y=3x-5y$

Solve for *a*.

13. $2(x+a)=x-4a$

14. $3(2a-5x)=2(a+3x)$

15. $3(2x+5)=\dfrac{1}{5}(x+a)$

16. The formula for the area of a triangle is $A = \dfrac{1}{2}bh$. Solve for h.

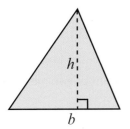

17. The formula for simple interest is $I = prt$. Solve for r.

18. A distance formula from physics is $s = \dfrac{1}{2}gt^2$. Solve for g.

19. The formula for the area of a trapezoid is $A = \dfrac{1}{2}h(b+c)$. Solve for b.

20. A formula for changing temperature from degrees Celsius to degrees Fahrenheit is $F = \dfrac{9}{5}C + 32°$. Solve for C.

21. A formula from physics is $s = h + vt + \dfrac{1}{2}gt^2$. Solve for v.

2-5   FIRST-DEGREE INEQUALITIES

OBJECTIVES

Upon completion of this section you should be able to:
1. Use the inequality symbol to represent the relative positions of two numbers on the number line.
2. Graph inequalities on the number line.
3. Solve first-degree inequalities.

Given any two real numbers a and b, it is always possible to state $a = b$ or $a \neq b$ (read "a is not equal to b"). Many times we are interested only in whether or not two numbers are equal, but there are situations in which we also wish to represent the relative sizes of numbers that are not equal.

The symbols $<$ and $>$ are **inequality symbols** or **order relations** and are used to show the relative sizes of the values of two numbers. The symbol $<$ is read "less than". For instance, $a < b$ is read "a is less than b". The symbol $>$ is read "greater than". For instance, $a > b$ is read "a is greater than b". The statement "a is less than b", $a < b$, is the same as saying "b is greater than a", $b > a$. One way to remember the meaning of the symbols is that the pointed end is toward the *lesser* of the two numbers.

DEFINITION

$a < b$ means that a is to the left of b on the real number line.
$a > b$ means that a is to the right of b on the real number line.

In the following examples we will locate the numbers on the number line and replace the ? with $<$ or $>$.

Example 1   3 ? 6

Solution

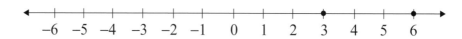

$3 < 6$ because 3 is to the left of 6. ∎

Example 2   −4 ? 0

Solution

$-4 < 0$ because −4 is to the left of 0. ∎

Example 3   4 ? –2

Solution

4 > –2 because 4 is to the right of –2. ■

Example 4   –6 ? –2

Solution

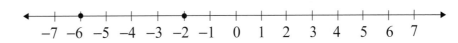

–6 < –2 since –6 is to the left of –2. ■

EXERCISE 2-5-1

Locate the following numbers on the number line and replace the question mark with > or <.

1. 6 ? 10

2. –6 ? –10

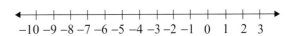

3. –3 ? 3

4. –4 ? –1

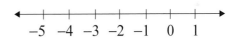

5. 4 ? 1

6. 1 ? 4

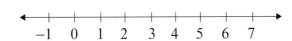

7. –2 ? –3

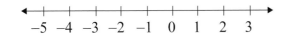

8. –5 ? –3

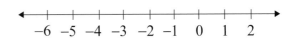

9. 0 ? 7

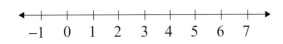

10. 0 ? –3

The mathematical statement $x < 3$ (read "x is less than 3") indicates that the variable x can be any number less than (or to the left of) 3. Remember, we are considering the real numbers and not just integers, so do not think of the values of x for $x < 3$ as only 2, 1, 0, −1, and so on.

As a matter of fact, to name the number x that is the largest number less than 3 is an impossible task. It can be indicated on the number line, however. To do this we need a symbol to represent the meaning of a statement such as $x < 3$.

> **DEFINITION**
> The left and right parentheses, when used on the number line, indicate the endpoint is *not* included in the set.

Example 5   Graph: $x < 3$

Solution

The graph in Example 5 is showing all real numbers less than 3. Note that the graph has an arrow indicating that the graph continues without end to the left.

Example 6   Graph: $x > -5$

Solution

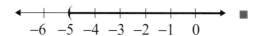

Example 7   Draw a number line graph showing that $x > -1$ and $x < 5$. The word "and" in a mathematical statement indicates both conditions must apply.

Solution

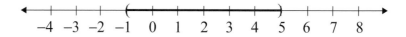

The statement, $x > -1$ and $x < 5$, can be condensed to read $-1 < x < 5$. ■

Example 8   Graph: $-3 < x < 3$

Solution

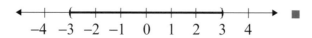

If we wish to include the endpoint in the set, we use a different symbol, $\leq$ or $\geq$. We read these symbols "less than or equal to" and "greater than or equal to".

Example 9 $x \geq 4$ indicates the number 4 *and* all real numbers to the right of 4 on the number line. ∎

> **DEFINITION**
>
> The left and right brackets, when used on the number line, indicate the endpoint *is* included in the set.

Example 10   Graph: $x \geq 1$

Solution

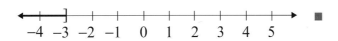

Example 11   Graph: $x \leq -3$

Solution

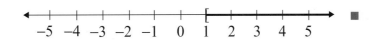

Example 12   Write an algebraic statement represented by the following graph.

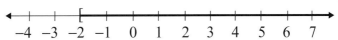

Solution $x \geq -2$ This can also be written as $-2 \leq x$. ∎

Example 13   Write an algebraic statement for the following graph.

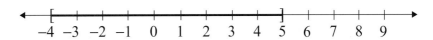

Solution $x \geq -4$ and $x \leq 5$ This can also be written as $-4 \leq x \leq 5$. ∎

Example 14   Write an algebraic statement for the following graph.

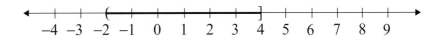

Solution $x > -2$ and $x \leq 4$ This can also be written as $-2 < x \leq 4$. ∎

Example 15   Graph: $x > 2\frac{1}{2}$

Solution   This example presents a small problem. How can we indicate $2\frac{1}{2}$ on the line? If we estimate the point, then another person might misread the statement. Could you possibly tell if the point represents $2\frac{1}{2}$ or maybe $2\frac{7}{16}$? Since the purpose of a graph is to clarify, always name the endpoint.

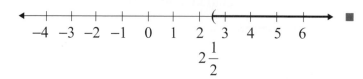

In summary, to graph a first-degree inequality, we ask the following questions.
1. Where is the starting point?
2. Is this point included or excluded?
3. Which direction is shaded?

EXERCISE 2-5-2

Construct a graph on the number line.

1. $x > 7$

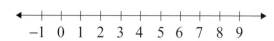

2. $x < 5$

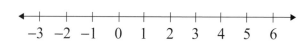

3. $x < -1$

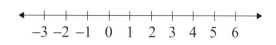

4. $x > -3$

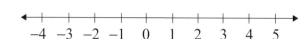

5. $x \geq 1$

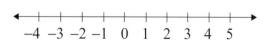

6. $x \geq -4$

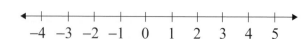

7. $3 < x < 7$

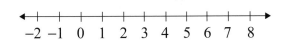

8. $-5 \leq x < 0$

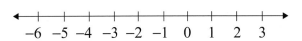

9. $x < 5.4$

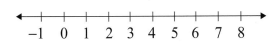

10. $6 < x \leq 7$

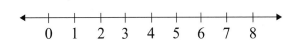

Write an algebraic statement for each graph.

11.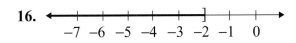

12.

13.

14.

15.

16.

17.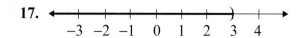

18.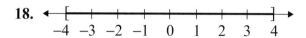

19.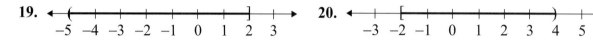

20.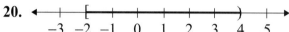

The solutions for first-degree inequalities involve the same basic rules as equations but with one exception.

RULE

If an inequality is multiplied or divided by a *negative* number, the inequality symbol is reversed.

To illustrate the above rule, consider the inequality $-5 < 8$.

Multiply both sides by -1.

$$(-1)(-5) \ ? \ (-1)(8)$$

$$5 \ ? \ -8$$

For the resulting statement to be true, we must insert a "greater than" symbol.

$$5 > -8$$

Example 16   Solve and graph: $-2x > 6$

Solution   To isolate x on the left side we must divide by -2.

$$\frac{-2x}{-2} < \frac{6}{-2} \qquad \text{Note the change from > to <.}$$

$$x < -3$$

CAUTION   Each time you multiply or divide both sides of an inequality by a *negative* number, you must change the direction of the inequality symbol. This is the *only* difference between solving first-degree equations and inequalities. Division or multiplication by a positive number will *not* change the direction of the inequality symbol.

The first four steps of the following procedure for solving inequalities are identical to those for solving equations. Only in step 5 is there a change in the procedure.

A PROCEDURE FOR SOLVING FIRST-DEGREE INEQUALITIES

1. Clear grouping symbols using the appropriate operation.
2. Multiply both sides by the least common denominator of all fractions appearing in the inequality.
3. Combine similar terms on each side of the inequality.
4. Add or subtract terms on both sides of the inequality to get the unknown on one side and everything else on the other.
5. Divide both sides of the inequality by the coefficient of the unknown. If the coefficient is positive, the inequality symbol will remain the same. If the coefficient is negative, the inequality symbol will be reversed.
6. Simplify the solution.

Example 17   Solve and graph: $2(x+2) < x-5$

Solution

$$2x+4 < x-5$$
$$2x+4-4 < x-5-4$$
$$2x < x-9$$
$$2x-x < x-9-x$$
$$x < -9$$

Example 18   Solve and graph: $2(x-3) \geq 3x-4$

Solution

$$2x - 6 \geq 3x - 4$$

$$2x - 6 + 6 \geq 3x - 4 + 6$$

$$2x \geq 3x + 2$$

$$2x - 3x \geq 3x + 2 - 3x$$

$$-x \geq 2$$

$$\frac{-x}{-1} \leq \frac{2}{-1}$$

$$x \leq -2$$

Example 19   Solve and graph: $\dfrac{2}{3}x + 2 > 3(x+1)$

Solution

$$\frac{2}{3}x + 2 > 3x + 3$$

$$3\left(\frac{2}{3}x + 2\right) > 3(3x + 3)$$

$$2x + 6 > 9x + 9$$

$$2x + 6 - 9x > 9x + 9 - 9x$$

$$-7x + 6 > 9$$

$$-7x > 3$$

$$\frac{-7x}{-7} < \frac{3}{-7}$$

$$x < -\frac{3}{7}$$

EXERCISE 2-5-3

Solve each inequality and graph.

1. $x + 3 < 7$

←—+——+——+——+——+——+——+——+——+——+——+——+——→

2. $x - 5 < 0$

←—+——+——+——+——+——+——+——+——+——+——+——+——→

3. $3x + 4 < 2x - 1$

←—+——+——+——+——+——+——+——+——+——+——+——+——→

4. $5x < 4x + 1$

←—+——+——+——+——+——+——+——+——+——+——+——+——→

5. $-3x > 9$

←—+——+——+——+——+——+——+——+——+——+——+——+——→

6. $-5x < -10$

7. $\frac{1}{2}x < 2$

8. $\frac{1}{5}x \leq -1$

9. $2(x+3) \leq x+9$

10. $5x-1 > 3(x+1)$

11. $-3(x-1) \geq 2(1-2x)$

12. $5(x+3) > 2(2x-1)$

13. $2(x+3) \leq 7(x+2)+2$

14. $\dfrac{2}{3}x+1 \geq 2x - \left(\dfrac{x}{2}-6\right)$

Example 2   Solve: $|4x - 3| = 10$

Solution

$$4x - 3 = -10 \quad \text{or} \quad 4x - 3 = 10$$

$$4x = -7 \quad \text{or} \quad 4x = 13$$

$$x = \frac{-7}{4} \quad \text{or} \quad x = \frac{13}{4} \quad \blacksquare$$

EXERCISE 2-6-1

Solve.

1. $|x| = 3$ **2.** $|x| = 0$ **3.** $|x + 5| = 8$ **4.** $|x + 7| = 2$

5. $|x - 6| = 5$ **6.** $|x - 3| = 0$ **7.** $|2x| = 6$ **8.** $|2x + 1| = 5$

9. $|3x - 2| = 7$ **10.** $|2x - 5| = 2$

Absolute value can occur in inequalities. For instance, the statement $|x| < 5$ means the number x is within five units of zero on the number line, so $x > -5$ *and* $x < 5$. Thus $|x| < 5$ is equivalent to saying that x is between -5 and 5 on the number line or $-5 < x < 5$. Therefore, the graph of $|x| < 5$ is

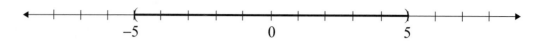

RULE

If a is a positive number, $|x| < a$ is equivalent to

$$-a < x < a.$$

Example 3   Solve and graph: $|x + 3| < 7$

Solution $|x + 3| < 7$ is equivalent to the statement

$$-7 < x + 3 < 7.$$

Using the rules for solving inequalities, we subtract 3 from each part of the inequality.

$$-10 < x < 4$$

The graph of this statement is

Example 4   Solve and graph: $\left| \dfrac{2x-1}{3} \right| < 5$

Solution
$$-5 < \frac{2x-1}{3} < 5$$
$$-15 < 2x - 1 < 15$$
$$-14 < \ 2x \ \ < 16$$
$$-7 < \ \ \ x \ \ \ < 8$$

The graph of this statement is

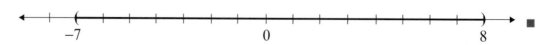

Similarly, $|x| > 5$ means that the real number x is more than five units from zero on the number line. Hence $|x| > 5$ is equivalent to $x < -5 \, or \, x > 5$. The graph of $|x| > 5$ is

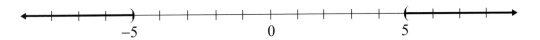

RULE

If a is a positive number, $|x| > a$ is equivalent to

$$x < -a \text{ or } x > a.$$

Example 5   Solve and graph: $|x + 3| \geq 7$

Solution $|x + 3| \geq 7$ is equivalent to the statement

$$x + 3 \leq -7 \text{ or } x + 3 \geq 7.$$

Solving these inequalities we find

$$x \leq -10 \text{ or } x \geq 4.$$

The graph of this statement is

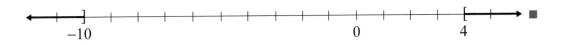

Example 6   Solve and graph: $|2x - 1| > 4$

Solution
$$2x - 1 < -4 \quad \text{or} \quad 2x - 1 > 4$$
$$2x < -3 \quad \text{or} \quad 2x > 5$$
$$x < \frac{-3}{2} \quad \text{or} \quad x > \frac{5}{2}$$

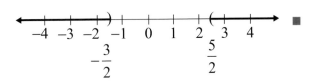

CAUTION   Take careful note of the fact that an *and* statement, such as $x > -5$ and $x < 5$, can be condensed to $-5 < x < 5$. However, an *or* statement, such as $x > 5$ or $x < -5$, cannot be condensed to read $-5 > x > 5$ since this implies -5 is greater than 5.

EXERCISE 2-6-2
Solve and graph.

1. $|x| < 3$

2. $|x| > 4$

3. $|x + 5| < 8$

4. $|x + 1| > 3$

5. $|x - 2| \leq 5$

6. $|x - 4| \geq 6$

7. $|2x+1| < 3$

8. $|3x-2| \geq 5$

9. $|4x-3| > 1$

10. $|x+5| \leq 7$

11. $\left|\dfrac{3x-1}{5}\right| \geq 1$

12. $\left|\dfrac{4x+3}{5}\right| \leq 3$

CHAPTER 2 SUMMARY

The number in brackets refers to the section of the chapter that discusses the concept.

Terminology

- An **equation** states that two numerical expressions are equal. [2–1]
- A **solution** to an equation is a value of the variable that makes the equation true. [2–1]
- Equations are **equivalent** if their solutions are identical. [2–1]
- A **literal equation** is an equation having more than one letter. [2–4]
- **Inequality symbols** or **order relations** are used to show the relative sizes of the values of two numbers. [2–5]
- The **absolute value** of a number is its distance from zero on the number line. [2–6]

$$|x| = \begin{cases} x & \text{if } x \geq 0 \\ -x & \text{if } x < 0 \end{cases}$$

Rules and Procedures

Solving First-Degree Equations

- If any number is added to or subtracted from both sides of an equation, the resulting equation will be equivalent to the original equation. [2–2]

- If both sides of an equation are multiplied or divided by the same nonzero number, the resulting equation is equivalent to the original equation. [2–2]

- To solve a first-degree equation, follow these steps: [2–3]
  1. Remove any parentheses.
  2. Multiply both sides by the least common denominator of all fractions appearing in the equation.
  3. Combine similar terms on each side of the equation.
  4. Add or subtract terms on both sides of the equation to get the unknown on one side and everything else on the other.
  5. Divide both sides of the equation by the coefficient of the unknown.
  6. Simplify the solution.

Inequalities

- If an inequality is multiplied or divided by a negative number, the inequality symbol is reversed. [2–5]

- To solve an inequality, follow these steps: [2–5]
  1. Remove any parentheses.
  2. Multiply both sides by the least common denominator of all fractions appearing in the inequality.
  3. Combine similar terms on each side of the inequality.
  4. Add or subtract terms on both sides of the inequality to get the unknown on one side and everything else on the other.
  5. Divide both sides of the inequality by the coefficient of the unknown. If the coefficient is positive, the inequality symbol will remain the same. If the coefficient is negative, the inequality symbol will be reversed.
  6. Simplify the solution.

- To graph a first-degree inequality, we ask the following questions. [2–5]
  1. Where is the starting point?
  2. Is this point included or excluded?
  3. Which direction is shaded?

Absolute Value Equations and Inequalities [2–6]

   If a is a positive number,

- $|x| = a$ is equivalent to $x = -a$ or $x = a$.

- $|x| < a$ is equivalent to $-a < x < a$.

- $|x| > a$ is equivalent to $x < -a$ or $x > a$.

CHAPTER 2 REVIEW

Determine if the following pairs of equations are equivalent.

1. $x + 3 = 2x - 1$ and $x = 4$ **2.** $3x - 5 = 2x + 7$ and $x = 2$ **3.** $3(x - 2) = 2x - 5$ and $x = -3$

Solve.

4. $x - 6 = -11$ **5.** $5x + 3 = 4x - 7$ **6.** $27y = 3$

7. $-\dfrac{y}{5} = 15$ **8.** $\dfrac{2}{3}x = \dfrac{3}{4}$ **9.** $2x + 3 = 3(2x + 5)$

10. $2(x + 5) = 6 + 4(x - 3)$ **11.** $\dfrac{x}{2} + \dfrac{2x}{3} = 7$ **12.** $3q - 5(q - 2) = \dfrac{1}{3}(q + 2)$

13. $\dfrac{2}{3}p - \dfrac{p}{2} + \dfrac{1}{4} = \dfrac{2}{3}(p - 4)$

Solve for the indicated variable.

14. $5 - 3x = a + 7x$ for x

15. $3(x - 2a) + 3a = 5(2x + a)$ for x

16. $4(3x - 2) = \dfrac{1}{3}(x - a)$ for a

17. $P = 2x + 2y$ for y

18. $H = \dfrac{yz + w}{xy}$ for z

19. $|x - 4| = 9$

20. $|3x + 1| = 4$

Solve and graph.

21. $3x \leq 15$

22. $-\dfrac{1}{2}x \leq 4$

23. $3x - 5 \leq x - 11$

24. $2(x - 5) + 3 > 3x - 1$

7. $x - \dfrac{1}{3}x - 2 = \dfrac{2}{5}(x+3)$

7. _____

8. $3x + 5y = 6$

8. _____

9. $\dfrac{2}{7}(x+a) = 4(2x-a)$

9. _____

10. $|3x+1| = 10$

10. _____

Graph.

11. $x \le 5$

12. $-1 \le x < 4$

13. $x < -3$ or $x > 0$

Solve for x.

14. $13 - x < 13$ **14.** _____

15. $\dfrac{2}{3}(x-1) - \dfrac{3}{4} \geq \dfrac{1}{2}(x+2)$ **15.** _____

16. $|5x - 11| \leq 19$ **16.** _____

17. $|3x + 2| > 11$ **17.** _____

18. The length of a certain rectangle is $12\dfrac{1}{2}$ cm. If the perimeter of the rectangle is 35 cm, find the width. Use the formula $P = 2l + 2w$.   **18.** _____

19. If $A = \dfrac{1}{2}h(a+b)$, solve for b.   **19.** _____

20. Solve for x and graph the results on the number line: $\dfrac{1}{2}x + 2 \geq x - \dfrac{1}{3}$ **20.** _____

CHAPTER 3

SURVEY

The following questions refer to material discussed in this chapter. Work as many problems as you can and check your answers with the answer key in the back of the book. The results will direct you to the sections of the chapter in which you need to work. If you answer all questions correctly, you may already have a good understanding of the material contained in this chapter.

1. Write an algebraic expression for each phrase.

   a. Twice the width, decreased by nine   1. a. _____

   b. One year more than one-third Vickie's age   b. _____

   c. 17.3% of the population   c. _____

   d. The value, in cents, of d dimes   d. _____

2. The second side of a triangle is twice the first, and the third side is two yards more than the first. If the perimeter of the triangle is 26 yards, find the length of each side.

   2. _____

3. A beginning tennis class has twice as many students as the advanced class. The intermediate class has three more students than the advanced class. How many students are in the advanced class if the total enrollment for the three tennis classes is 43?

   3. _____

4. Salt Lake City and Butte are 677 kilometers apart. Terry leaves Salt Lake City bound for Butte at 75 kilometers per hour. Twenty minutes later, Jim leaves Butte bound for Salt Lake City at 88 kilometers per hour. How many hours after Jim leaves Butte will they meet?

4. _____

5. The ratio of miles to kilometers is 5 to 8. How many miles are in 40 kilometers?

5. _____

6. How many liters of pure dye must be added to 45 liters of a 6% dye solution to obtain a 10% dye solution?

6. _____

Solving Word Problems

Solving problems of any type is a necessary but sometimes difficult skill to develop. In 1945, George Polya, a Hungarian mathematician (1887-1985), published a general method to help people solve problems. In his book, *How to Solve It* [Princeton University Press, Princeton, NJ], Polya discussed how to apply his method to the solution of many types of problems.

Polya believed that problem solving can be accomplished in four steps: understand the problem; devise a plan; carry out the plan; look back. There are many techniques, such as working backwards, trial and error, and creating a systematic list, that can help you solve problems. In this chapter, we will discuss using algebra to solve problems.

When solving problems algebraically, use Polya's method as follows:
Understand the problem. Read the problem carefully and think about it.
Devise a plan. Write variable expressions for the unknowns and an equation for the problem.
Carry out the plan. Solve the equation.
Look back. Be sure you have answered the question and that your answer makes sense.

3–1   FROM WORDS TO ALGEBRA

> **OBJECTIVES**
>
> Upon completion of this section you should be able to:
> 1. Change an English phrase into an algebraic expression.
> 2. Express a relationship between two or more unknowns in a given statement by using one variable.

Consider the following problem.

Dave is four years younger than Sara, and Sara is two years younger than Andrew. The sum of all their ages is 64. Find the age of each person.

After reading this problem, we discover that to carry out the plan of using algebra, we need to translate the words into algebraic symbols. We will solve this problem completely in Section 3–2. First we need to be able to translate from one language to another–from English to algebra. The process of translation is an important one, so we will provide some suggestions and examples for translation in this section.

The following table includes some common phrases and the corresponding algebraic expressions. We will let x represent "a number" in each instance.

Operation	English Phrase	Algebraic Expression
Addition	A number increased by seven	$x + 7$
	Three more than a number	$3 + x$ or $x + 3$
	The sum of a number and two	$x + 2$
Subtraction	The difference of a number and 3	$x - 3$
	The difference of eight and a number	$8 - x$
	A number decreased by seven	$x - 7$
	Four less than a number	$x - 4$
Multiplication	The product of a number and six	$6x$
	Four times a number	$4x$
	Twice a number	$2x$
	One fourth of a number	$\frac{1}{4}x$ or $\frac{x}{4}$
Division	The ratio of a number and three	$\frac{x}{3}$
	The quotient of two and a number	$\frac{2}{x}$

Although x is often used for an unknown quantity, when possible we will choose variables that are more representative of the quantity.

Example 1   Write an algebraic expression for the phrase:
The number of students, x, increased by three

Solution $x + 3$ ∎

Example 2   Write an algebraic expression for the phrase:
Three feet less than the length

Solution   We let *l* represent the length. Then the expression is $l - 3$. ■

Example 3   Write an algebraic expression for the phrase:
Three times the number of dollars

Solution   We let *d* represent the number of dollars. Then the expression is $3d$. ■

Example 4   Write an algebraic expression for the phrase:
Five percent of the population

Solution   We let *p* represent the population. Five percent is written as 0.05. So the expression is $0.05p$. ■

Example 5   Write an algebraic expression for the phrase:
The number of gallons in *q* quarts

Solution   There are 4 quarts in a gallon. The number of gallons is found by dividing *q* by 4. The expression is $\frac{q}{4}$. ■

Example 6   Write an algebraic expression for the phrase:
The number of days in *x* weeks and five days

Solution   There are 7 days in a week, so the number of days in *x* weeks is found by multiplying 7 and *x*. The expression is $7x + 5$. ■

Example 7   Write an algebraic expression for the phrase:
The value, in cents, of *d* dimes and seven pennies

Solution   The value of a dime is 10 cents. Therefore, to indicate the value of *d* dimes, multiply 10 by *d*, obtaining $10d$. The expression is $10d + 7$. ■

EXERCISE 3-1-1

Write an algebraic expression for each phrase.

1. The number of airplane tickets to Florida increased by 20

2. Three inches more than John's height

3. Double the amount of flour in a recipe

4. Twice as many hours worked

5. Four minutes less than Greg's time in a race

6. Becca's number of credits decreased by three

7. One-half the amount of saline solution

8. Two-thirds the height of the flagpole

9. Three feet less than twice the height of a triangle

10. Five kilometers more than three times the distance to school

11. 6% of a retirement investment

12. 10% interest on a credit card balance

13. The total value, in cents, of q quarters and two dimes

14. The total value, in cents, of n nickels and three quarters

15. The number of cups in x quarts of milk

16. The number of days in h hours

When it is necessary to find the values of more than one quantity in a problem, we will write a variable expression for each quantity. We begin by choosing a variable for one of the unknown quantities. Then we write expressions in terms of this variable for the other quantities.

Example 8   The length of a rectangle is three meters more than the width. Write expressions for the length and width using one variable.

Solution   Both the length and width are unknown quantities, so we must choose a variable to represent one of them. Then we will write an expression that represents the other using this variable. There are two ways to accomplish this.

First, we will let w represent the width.

Now, since we are told that the length is three meters more than the width, we can write the length of the rectangle as $w + 3$.

We can draw a sketch and label the sides as follows to indicate our choice of variable.

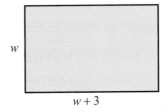

Suppose we had decided to let l represent the length of the rectangle. Now we would have to recognize that the statement "the length is three meters more than the width" is the same as saying "the width is three meters less than the length." Therefore, we would represent the width of the rectangle as $l - 3$ and label the rectangle as follows.

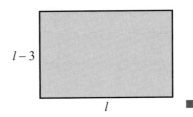

From this example we see that there can be more than one way to identify the unknowns. Therefore, it is extremely important that we designate which unknown is represented by each expression. This can be done in a statement such as "let $w =$ the width of the rectangle", by labeling a diagram when appropriate, or by filling in a table with the unknown quantities and the corresponding algebraic expressions. The following table would be appropriate for Example 8.

Unknown	Expression
Width of the rectangle	w
Length of the rectangle	$w + 3$

Example 9   The Sears Tower in Chicago is eight stories taller than the Empire State Building in New York. Write algebraic expressions for the number of stories in each building using one variable.

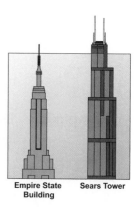

Empire State   Sears Tower
Building

Solution   From the information given, we do not know how many stories either building has. We will choose one of the buildings and represent the number of stories by x.
Let x = the number of stories in the Empire State Building.
The information states that the Sears Tower is eight stories taller than the Empire State Building. So we write $x + 8$ = the number of stories in the Sears Tower.

Unknown	Expression
Number of stories in the Empire State Building	x
Number of stories in the Sears Tower	$x + 8$

Example 10   The width of a rectangle is one-fourth the length. Write expressions for the length and width using one variable.

Solution   We will represent the length using l. Then the width can be represented by $\dfrac{l}{4}$.

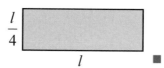

Example 11   The sum of two numbers is 20. Write expressions for both numbers using one variable.

Solution   We have two numbers that add to 20. Suppose you choose a number, say 4, for the first number. What is the second number? The second number would be 16 because $4 + 16 = 20$. How did you get 16? Most likely you subtracted 4 from 20. So, if we let x represent the first number, the second number would be $20 - x$. ∎

The numerical technique used in Example 11 can be very helpful in writing algebraic expressions for unknown quantities. When presented with a new problem, try to work out a few possibilities numerically. Then write the variable expressions using a pattern that you find.

Example 12   Dave is four years younger than Sara, and Sara is two years younger than Andrew. Write expressions for their ages.

Solution   We need to choose a variable for one of the ages and then write expressions for the other two ages in terms of this variable. One way to do this is to let *x* represent Sara's age. Then we have the following.

Unknown	Expression
Sara's age	x
Dave's age	$x - 4$
Andrew's age	$x + 2$

■

EXERCISE 3-1-2

Express the unknown quantities as variable expressions. Each problem has more than one correct answer.

1. Becky is six years older than Zack. Write expressions for their ages.

2. During the second week of production of a new car, General Motors produced 612 units more than it did during the first week. Write expressions for the number of cars produced in each week.

3. A 9:00 A.M. math class has eight fewer students than an 11:00 A.M. class. Write expressions for the number of students in each class.

4. The population of Charleston, SC is four times the population of Johnstown, PA. Write expressions for the population of each city.

5. The length of a rectangular garden is five meters more than twice the width. Write expressions for the length and width.

6. The length of a credit card is three centimeters more than the width. Write expressions for the length and width.

7. Emily makes two investments that total $10,000. Write expressions to represent the two investments.

8. Two beakers contain a total of 800 ml of a saline solution. Write expressions for the amount of saline solution in each beaker.

9. Last year's enrollment at a college was 7% less than this year's. Write expressions for the enrollment for last year and this year.

10. The biochemistry class has four more students than the physics class. The biochemistry class has nine fewer students than the algebra class. Write expressions for the number of students in each class.

11. A recipe for brownies uses half as much flour as a recipe for pancakes. The recipe for brownies uses three times as much flour as a recipe for crepes. Write expressions for the amount of flour in each recipe.

12. Jackie has twice as many dimes as nickels and two more quarters than dimes. Write expressions for the number of each kind of coin Jackie has.

13. Anita has five dollars more than Ryan, and she has thirteen dollars less than Lucas. Write expressions for the number of dollars each person has.

14. A Chrysler Sebring obtains eight miles per gallon less than a Chevrolet Cavalier. The Sebring gets five miles per gallon more than a Subaru Forester. Write expressions for the gas mileage of the three cars.

15. A certain amount is invested in an account for one year at 5% interest. Write expressions for the original amount, the interest earned, and the total amount in the account at the end of the year.

3–2   AN INTRODUCTION TO SOLVING WORD PROBLEMS

> **OBJECTIVE**
>
> Upon completion of this section you should be able to use algebra to solve word problems.

In the previous section we presented examples and exercises designed to help you learn the skill of translating English phrases into algebraic expressions. In this section we will complete the problem solving process. The next step in the solution of a word problem is the translation of the problem into an algebraic *equation*. We look for two quantities described or implied as being equal. We set the expressions for these quantities equal to one another to create the equation. Then we solve the equation, and finally we look back to make sure this solution fits the problem. This completes the algebraic method of solving word problems.

Example 1   The Sears Tower in Chicago is eight stories taller than the Empire State Building in New York. The total number of stories in both buildings is 212. Find the number of stories in each building.

Solution   The first sentence in this problem is the same as Example 9 from the previous section. There x represented the number of stories in the Empire State Building and $x + 8$ represented the number of stories in the Sears Tower.

Now we need to find two equal quantities. The problem states that the *total* number of stories *is* 212. The word "total" we translate as addition, and the word "is" we translate as equals. This gives

$$\overbrace{x+(x+8)}^{\text{total \# of stories}} \overbrace{=}^{\text{is}} \overbrace{212}^{212}$$

Solve the equation.
$$2x + 8 = 212$$
$$2x = 204$$
$$x = 102$$

Now that we have a value for x, we need to look back to see if we solved the problem. The value of x only gives the number of stories in the Empire State Building. We also need to find the number of stories in the Sears Tower. So we evaluate the expression $x + 8$ for $x = 102$ to get 110. Finally, we write the solution.

The number of stories in the Empire State Building is 102.

The number of stories in the Sears Tower is 110. ∎

The remainder of this section contains examples of solving word problems involving percent, ratios, geometry, and numerical relationships. We are presenting a variety of examples so that you begin to appreciate that this method can be applied in many situations.

Remember, the method is:

Understand the problem. Read the problem carefully and think about it.

Devise a plan. Write variable expressions for the unknowns and an equation for the problem.

Carry out the plan. Solve the equation.

Look back.   Be sure you have answered the question and that your answer makes sense.

Example 2   The length of a rectangle is three meters more than the width. Find the length and width if the perimeter of the rectangle is 26 meters.

Solution   In Example 8 in the previous section we wrote the expressions for the length and width in two ways. To see that each of these translations will result in a correct solution, we will write an equation for each translation and solve.

First, let w represent the width. Then $w + 3$ is the length.

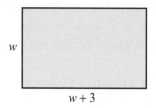

Recall that the formula for perimeter is $P = 2w + 2l$. We can use the fact that the perimeter is equal to 26 meters to write an equation for this problem.

$$26 = 2w + 2(w + 3)$$

Solve this equation.

$$26 = 2w + 2w + 6$$
$$26 = 4w + 6$$
$$20 = 4w$$
$$5 = w \quad \text{or} \quad w = 5$$

Therefore, the width is 5 meters and the length is 3 meters more or 8 meters.

As an alternative, let l represent the length and $l - 3$ represent the width.

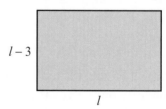

The equation becomes

$$26 = 2(l - 3) + 2l.$$

Solve.

$$26 = 2l - 6 + 2l$$
$$26 = 4l - 6$$
$$32 = 4l$$
$$8 = l$$

The length is 8 meters and the width is 5 meters. ∎

CAUTION   Remember to perform the last step in the problem solving process. Do not just solve for the variable and think you have solved the problem. *Look back* by reading the problem one more time to be sure you have answered the question(s).

Example 3   Dave is four years younger than Sara, and Sara is two years younger than Andrew. The sum of all their ages is 64. Find the age of each person.

Solution   From Example 12 in Section 3-1, we have the following expressions.

Unknown	Expression
Sara's age	x
Dave's age	$x - 4$
Andrew's age	$x + 2$

The equation can be written by translating the second sentence in the problem as follows.

$$\overbrace{x+(x-4)+(x+2)}^{\text{The sum of their ages}} \overbrace{=}^{\text{is}} \overbrace{64}^{64.}$$

Solve.
$$3x - 2 = 64$$
$$3x = 66$$
$$x = 22$$

This gives Sara's age as 22. Dave's age is 18 and Andrew's age is 24. ∎

Example 4   A length of rope is 64 meters. It is cut into three pieces. The second piece is four meters shorter than the first. The first piece is two meters shorter than the third. Find the length of each piece.

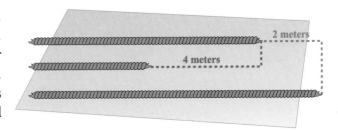

Solution   You might at first think that this is a new problem. However, if we compare this problem with Example 3, we see that the relationship of the lengths of the pieces of rope is the same as the relationship of the ages. Let x represent the length of the first piece of rope and write the expressions for the remaining pieces as in the table.

Unknown	Expression
Length of first piece	x
Length of second piece	$x - 4$
Length of third piece	$x + 2$

The entire length is 64 meters. So the equation is

$$x + (x-4) + (x+2) = 64$$

Solving, we have $x = 22$. Therefore, the first piece is 22 meters long, the second piece is 18 meters long, and the third piece is 24 meters long. ∎

Example 5   Katie purchases a mountain bike for \$667.80, including a 6% sales tax. Find the retail price of the bike and the amount of sales tax.

Solution   The total purchase price of the bike includes the retail price and the sales tax. Let x represent the retail price of the bike. The sales tax is found by multiplying the retail price of the bike by 6% or 0.06.

$$\underbrace{x}_{\text{retail price of bike}} + \underbrace{0.06x}_{\text{sales tax}} = \underbrace{667.80}_{\text{total price}}$$

Solve.

$$1.06x = 667.80$$
$$x = 630$$

The price of the bike is \$630 and the sales tax is 0.06(\$630) = \$37.80. ∎

Example 6   Jim wins a lottery prize of \$8000 and invests it for one year. He invests part at 6% and the rest at 5%. If the total simple interest at the end of the year is \$460, how much was invested at each rate?

Solution   Here we will use the simple interest formula $I = prt$ where I represents the interest, p the principal (amount invested), r the rate of interest, and t the time in years.
Let x = amount invested at 5% and $8000 - x$ = amount invested at 6%.

Then the interest earned at 5% is $I = prt = (x)(0.05)(1) = 0.05x$.

The interest earned at 6% is $I = prt = (8000 - x)(0.06)(1) = 0.06(8000 - x)$.

The equation representing the total interest is

$$\underbrace{0.05x}_{\text{interest earned at 5%}} + \underbrace{0.06(8000 - x)}_{\text{interest earned at 6%}} = \underbrace{460.}_{\text{total interest}}$$

Solve.

$$0.05x + 480 - 0.06x = 460$$
$$-0.01x + 480 = 460$$
$$-0.01x = -20$$
$$x = 2000$$
$$8000 - x = 6000$$

Thus \$2000 was invested at 5% and \$6000 was invested at 6%. ∎

Example 7   In a rectangle the length is two inches more than the width. A new rectangle is formed when the length is increased by three inches and the width is increased by one inch. The perimeter of the new rectangle will be twice that of the original rectangle. Find the length and width of the original rectangle.

Solution   Let w represent the width of the original rectangle.

Unknown	Expression
Width of original rectangle	w
Length of original rectangle	$w + 2$
Width of new rectangle	$w + 1$
Length of new rectangle	$(w + 2) + 3$ or $w + 5$

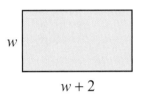

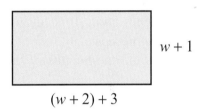

Remember that the perimeter of a rectangle can be found using the formula $P = 2w + 2l$. Since we are comparing two rectangles, we will need to use the formula twice.

$$\overbrace{2(w+5)+2(w+1)}^{\text{perimeter of new rectangle}} \overset{\text{is}}{=} \overbrace{2\big[2w+2(w+2)\big]}^{\text{twice the perimeter of original}}$$

Solve.
$$2(w+5)+2(w+1) = 2\big[2w+2w+4\big]$$
$$2w+10+2w+2 = 2(4w+4)$$
$$4w+12 = 8w+8$$
$$-4w+12 = 8$$
$$-4w = -4$$
$$w = \frac{-4}{-4} = 1$$

The width of the original rectangle is 1 inch and the length is $1 + 2 = 3$ inches. ∎

A **proportion** is a statement that two ratios are equal. The next example illustrates how to use a proportion as the equation in solving a problem.

Example 8   The ratio of miles to kilometers is approximately 5 to 8. Determine the number of miles in 100 kilometers.

Solution   Let x represent the number of miles in 100 kilometers. One ratio of miles to kilometers is $\dfrac{5}{8}$. The second ratio of miles to kilometers is $\dfrac{x}{100}$.

The proportion is $\dfrac{5}{8} = \dfrac{x}{100}$.

Solve. $100\left(\dfrac{5}{8}\right) = 100\left(\dfrac{x}{100}\right)$

$62.5 = x$

There are approximately 62.5 miles in 100 kilometers. ■

It is possible (and most likely probable) that you can solve some of these exercises using problem solving techniques other than algebra. However, we encourage you to use the algebraic method for all the problems so that you become proficient in its use.

EXERCISE 3-2-1

Solve each of the following problems.

1. A board is 22 inches long. It is cut into two pieces so that one piece is eight inches longer than the other. Find the lengths of the two pieces.

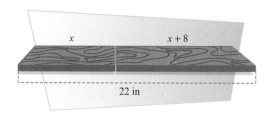

2. During the second week of production of a new car, General Motors produced 612 units more than it did during the first week. If the total production for the two weeks was 18,416 cars, find the production for each week. (See Exercise 3-1-2, #2.)

3. A 9:00 A.M. math class has eight fewer students than a 11:00 A.M. class. Find the number of students in each class if the total number of students in both classes is 84. (See Exercise 3-1-2, #3.)

4. The population of Charleston, SC is four times the population of Johnstown, PA. The total population of the two cities is 120,000. Find the population of each city. (See Exercise 3-1-2, #4.)

19. The base of a triangle is ten centimeters. If the height is increased so that it is four centimeters more than twice the original height, and the base is decreased by two centimeters, then the area of the new triangle is twice the area of the original. Find the height of the original triangle.

(Hint: Use $A = \dfrac{1}{2}bh.$)

20. The ratio of kilograms to pounds is 5 to 11. How many kilograms are in 121 pounds?

21. Nicholas weighs five kilograms less than Matthew, and Matthew weighs two kilograms less than Adam. Find the weight of each person if their total weight is 222 kilograms.

22. The ratio of the teeth on gear C to those on gear D is 7 to 9. If gear D has 36 teeth, how many teeth does gear C have?

23. Aaron's lawnmower calls for a ratio of oil to gas of 1 to 32. How many ounces of oil should he add to 2 gallons of gas? (Note: 1 gallon = 128 ounces)

24. Water is mixed with orange juice concentrate in a ratio of 3 to 1. How many ounces of water should be mixed with a 12-ounce can of orange juice concentrate? How many ounces of orange juice will this make?

3–3   DISTANCE-RATE-TIME PROBLEMS

> **OBJECTIVE**
>
> Upon completion of this section you should be able to solve distance-rate-time problems.

The problems in this section all concern objects that are in motion. When an object is moving at a constant or average rate for a given time, the distance it has traveled is found by the formula

$$d = rt \; (\text{distance} = \text{rate} \times \text{time}).$$

Example 1   Chris drove on the interstate at an average rate of 60 miles per hour for three hours. How far did he travel?

Solution   Since we know the rate and time, we substitute these values into the formula.

$$d = \left(60 \; \frac{\text{miles}}{\text{hour}}\right)(3 \text{ hours}) = 180 \text{ miles}$$

Chris drove 180 miles.   ■

Example 2   How long will it take Laura to ride her bike 60 miles if her average speed is 15 miles per hour?

Solution   Here we know distance and rate. Substitute and solve.

$$60 \text{ miles} = \left(15 \; \frac{\text{miles}}{\text{hour}}\right)(t \text{ hours})$$

$$60 = 15t$$

$$4 = t$$

Laura can ride 60 miles in 4 hours.   ■

The above examples illustrate how to use the formula $d = rt$ to solve problems with one object in motion. We can also use this formula to solve problems with two objects in motion. However, we will need to apply the formula twice. To organize the information in problems with two objects in motion, we will use a table.

Example 3   A thief leaves town heading north at an average speed of 120 kilometers per hour. The sheriff leaves two hours later in a plane that travels at 200 kilometers per hour. How long will it take the sheriff to catch the thief?

Solution   We start by organizing the given information. The problem gives a rate for each person and a comparison of their times. We are asked to find the time the sheriff will need to catch the thief, so we let t represent this time. The thief had a 2-hour head start, so his time is $t + 2$.

	r	$\cdot$ t	$=$ d
Thief	120	$t + 2$	$120(t + 2)$
Sheriff	200	t	$200t$

Use the formula $d = rt$ to find expressions for the distance each traveled.

The thief and sheriff start at the *same point* and travel in the *same direction*.

Thief
Sheriff

Therefore, when the sheriff catches the thief, they will have traveled equal distances.

$$\overbrace{120(t + 2)}^{\text{thief's distance}} = \overbrace{200t}^{\text{sheriff's distance}}$$
$$120t + 240 = 200t$$
$$240 = 80t$$
$$3 = t$$

Three hours after the sheriff leaves, he will catch the thief. ∎

Example 4   Pamela and Kristen start at the same point and go in opposite directions. The rate at which they are walking away from each other is 7 miles per hour. At the end of two hours Kristen stops, and Pamela continues to walk for another hour. When Pamela stops, she has walked twice as far as Kristen. Find the rate of each. How far apart are they when Kristen stops?

Solution   First, we need to organize the given information. The problem gives us the time each person walks, a total for their rates, and a comparison of their distances. Kristen's time is 2 hours and Pamela's time is 3 hours. Since we are asked to find the rates, we let
$$r = \text{Pamela's rate.}$$
Then $7 - r = \text{Kristen's rate.}$
We enter this information in the table.

	r	$\cdot$ t $=$	d
Pamela	r	3	$3r$
Kristen	$7 - r$	2	$2(7 - r)$

Multiply the rate and time of each to find an expression for distance.

Pamela and Kristen start at the *same point* and travel in *opposite directions*.

Pamela   Kristen

To write the equation, we use the fact that Pamela's distance is twice Kristen's distance.

$$\overbrace{3r}^{\text{Pamela's distance}} = \overbrace{2\big[2(7-r)\big]}^{\text{twice Kristen's distance}}$$

$$3r = 2\big[14 - 2r\big]$$
$$3r = 28 - 4r$$
$$7r = 28$$
$$r = 4$$

Pamela walks at 4 miles per hour and Kristen walks at $7 - 4 = 3$ miles per hour.
To find the distance between Pamela and Kristen, we find the distance of each and add.

  Pamela walked (3 hours)(4 mph) = 12 miles.
  Kristen walked (2 hours)(3 mph) = 6 miles.

They are 18 miles apart. ∎

Example 5   Juan and Steven started 6 kilometers apart and walked toward each other, meeting in 30 minutes. If Juan walked two kilometers per hour faster than Steven, find the rate of each. How far did each person walk?

Solution   We have the time Juan and Steven walked, a comparison of their rates, and the total distance. Since the rates are given in kilometers per hour and the time in minutes, we must rewrite 30 minutes as $\dfrac{1}{2}$ hour.

Let $r =$ Steven's rate.
Then $r + 2 =$ Juan's rate.

	r	$\cdot$ t $=$	d
Juan	$r + 2$	$\dfrac{1}{2}$	$\dfrac{1}{2}(r+2)$
Steven	r	$\dfrac{1}{2}$	$\dfrac{1}{2}r$

Multiply the rate and time of each to find an expression for distance.

Juan and Steven start at *different points* and travel *toward* each other.

Juan   Steven

Therefore, the sum of their distances is equal to the total distance between them.

$$\overbrace{\frac{1}{2}r}^{\text{Steven's distance}} + \overbrace{\frac{1}{2}(r+2)}^{\text{Juan's distance}} = \overbrace{6}^{\text{total distance}}$$

$$\frac{1}{2}r + \frac{1}{2}r + 1 = 6$$

$$r + 1 = 6$$

$$r = 5$$

Steven walks at a rate of 5 km/hr and Juan walks at a rate of 7 km/hr.

Steven walked $\left(\frac{1}{2}\ \text{hr}\right)(5\ \text{km/hr}) = 2\frac{1}{2}$ km.

Juan walked $\left(\frac{1}{2}\ \text{hr}\right)(7\ \text{km/hr}) = 3\frac{1}{2}$ km. ■

Example 6   An airplane, whose speed in still air is 550 miles per hour, flies against a headwind of 50 miles per hour. How long will it take to travel 1500 miles?

Solution   The speed of the plane in still air (550 mph) is reduced by the speed of the headwind (50 mph). We use the formula $d = rt$ to solve this problem.

$$d = rt$$
$$1500 = (550 - 50)t$$
$$1500 = 500t$$
$$t = 3$$

Thus it will take 3 hours to travel the distance. ■

Example 7   Mike can row his boat upstream from the hunting lodge to the park in five hours. He can row back downstream from the park to the lodge in three hours. If the stream is flowing at the rate of two kilometers per hour, how far is it from the lodge to the park?

Solution   Here we have one object in motion, the boat. However, we need to treat it as 2 objects since we assume that the rate of the stream will increase or decrease the rate of the boat by two kilometers per hour. When Mike is rowing upstream, the current will slow his progress. When Mike is rowing downstream, the current will add to his speed.
If we let
    x = Mike's rate in still water
then $x - 2$ = his upstream rate and $x + 2$ = his downstream rate.

	r	$\cdot\ t\ =$	d
Upstream	$x - 2$	5	$5(x-2)$
Downstream	$x + 2$	3	$3(x+2)$

Since Mike made a round trip to the park and back, the distance upstream is equal to the distance downstream.

$$\overbrace{5(x-2)}^{\text{distance upstream}} = \overbrace{3(x+2)}^{\text{distance downstream}}$$

$$5x - 10 = 3x + 6$$

$$2x - 10 = 6$$

$$2x = 16$$

$$x = 8$$

Notice that 8 is *not* the solution to the problem, but is Mike's rate of rowing in still water. However, the question asked is, "What is the distance from the lodge to the park?" To answer this, use either the distance upstream or the distance downstream since they are the same. Using the upstream distance, we have

$$5(x-2) = 5(8-2) = 30 \text{ km.} \quad \blacksquare$$

EXERCISE 3-3-1

1. A train travels at the rate of 88 miles per hour for three hours. How far has it traveled?

2. Matt walks at the rate of four miles per hour. How long will it take him to walk ten miles?

3. Amy rode her bike 57 miles in three hours. What was her average speed?

4. A van leaves Pittsburgh and travels west at the rate of 60 kilometers per hour. A truck leaves from the same point two hours later and travels the same route at 80 kilometers per hour. How long will it take the truck to catch the van?

	r	$\cdot\ t$	$=d$
Van	60	t	
Truck	80	$t-2$	

5. David and Julian leave the same point at 2:00 PM and ride their bikes in opposite directions. David travels at the rate of 20 miles per hour and Julian travels at 25 miles per hour. At what time will they be 90 miles apart?

6. Frank and Mike live 22 miles apart. If each leaves his home at the same time and walks toward the other, Frank at a rate of six miles per hour and Mike at five miles per hour, how long will it take for them to meet?

7. Keisha leaves Omaha and drives east at 70 kilometers per hour. Three hours later Alex leaves the same point and drives west at 85 kilometers per hour. How long will Alex have to drive until the two cars are 520 kilometers apart?

8. Race-car driver Jeff Gordon is maintaining an average speed of 184 miles per hour while Dale Earnhardt, Jr. is one-fifth mile behind him maintaining an average speed of 188 miles per hour. How long will it take Earnhardt, Jr. to catch Gordon?

9. A train leaves Grand Central Station for Boston and travels at the rate of 80 kilometers per hour. The express train leaves the same station one hour later and travels the same route at 120 kilometers per hour. How long after the express train leaves will the two trains be 60 kilometers apart?

10. A plane leaves New York bound for Houston traveling at the rate of 654 kilometers per hour. Thirty minutes later another plane leaves New York also bound for Houston traveling at the rate of 763 kilometers per hour. If both planes arrive at Houston at the same time, what is the distance between New York and Houston?

11. Chris starts at a certain point and walks due east. One hour later Beth starts at the same point and rides a bicycle due west, traveling nine miles per hour faster than Chris. After Beth has ridden for three hours they are 69 miles apart. What was the average rate of each?

	r	$\cdot$	t	$=$	d
Chris	r		4		
Beth	$r+9$		3		

12. Joan and Sally live five miles apart. They decide to meet for lunch in one-half hour at a restaurant that is located between them. They leave their homes at the same time, and Joan walks at a rate two miles per hour faster than Sally. How fast does each walk if they both arrive at the restaurant in exactly one-half hour?

13. Cheyenne and Rapid City are 300 miles apart. Nancy leaves Cheyenne at the same time Barb leaves Rapid City. They meet in two and one-half hours. If Barb traveled ten miles per hour faster than Nancy, what was the speed of each? How far from Cheyenne are they when they meet?

14. A stream is flowing at the rate of four kilometers per hour. A motorboat that can travel at thirty-two kilometers per hour in still water travels downstream. How long will it take to travel 99 kilometers?

15. A plane travels to Memphis with a 30 kilometer per hour tailwind in three hours and returns against the wind in three hours and thirty minutes. Find the speed of the plane in still air.

	r	$\cdot$ t $=$	d
To Memphis	$r+30$	3	
Return Trip	$r-30$	$\dfrac{7}{2}$	

16. A plane flew for $3\dfrac{1}{2}$ hours with a 30 mile per hour tailwind. It took five hours for the return trip. Find the speed of the plane in still air.

17. A motorboat travels upstream in five hours and makes the return trip downstream in three hours. If the rate of the stream's current is ten kilometers per hour, find the rate of the boat in still water.

18. A motorboat traveling at the rate of thirty kilometers per hour in still water takes four hours to go upstream and three hours to return. Find the rate of the current.

3–4   MIXTURE PROBLEMS

> **OBJECTIVE**
>
> Upon completion of this section you should be able to solve mixture problems.

The final type of problem we will discuss in this chapter is the mixture problem. Mixture problems come in various settings. They may involve mixing coffee, coins, liquids, and so on. Some people who solve mixture problems in their work are merchants, chemists, beauticians, manufacturers, pharmacists, and mechanics.

Example 1   A merchant has French Roast coffee that sells for $5.20 per pound and Colombian coffee that sells for $8.40 per pound. The merchant wishes to make a blend of the two coffees. How many pounds of each coffee must be used to make 20 pounds of a mixture that would sell for $6.00 per pound?

Solution   We use the fact that the merchant wishes to obtain 20 pounds of the mixture and let
x = the number of pounds of French Roast
and $20 - x$ = the number of pounds of Colombian.

Type	Pounds	· Price	= Total Value
French Roast	x	5.20	$5.20x$
Colombian	$20 - x$	8.40	$8.40(20 - x)$
Mixture	20	6.00	$6.00(20)$

Looking at the "total value" column of the table, notice that the sum of the total values of the two types of coffee must equal the total value of the mixture.

$$\overbrace{5.20x}^{\text{value of French Roast}} + \overbrace{8.40(20 - x)}^{\text{value of Colombian}} = \overbrace{6.00(20)}^{\text{value of mixture}}$$

$$5.20x + 168 - 8.40x = 120$$
$$-3.2x + 168 = 120$$
$$-3.2x = -48$$
$$x = 15$$

Thus the merchant must use 15 pounds of French Roast and 5 pounds of Colombian. ∎

Example 2

A total of 100 coins in nickels and quarters has a value of $14.40. How many of each type of coin are there?

Solution

We are given that there are 100 coins.
Let x = the number of nickels.
Then $100 - x$ = the number of quarters.
We must also know that nickels are valued at five cents each and quarters are valued at twenty-five cents each, and we must express these values as decimals. Again, a table is helpful.

Type	Number ·	Value	= Total Value
Nickels	x	0.05	$0.05x$
Quarters	$100 - x$	0.25	$0.25(100 - x)$
Mixture	100		14.40

The sum of the total value of nickels and the total value of quarters is equal to the total value of the mixture, $14.40.

$$\overbrace{0.05x}^{\text{value of nickels}} + \overbrace{0.25(100 - x)}^{\text{value of quarters}} = \overbrace{14.40}^{\text{total value of coins}}$$

$$0.05x + 25 - 0.25x = 14.40$$
$$-0.20x + 25 = 14.40$$
$$-0.20x = -10.6$$
$$x = 53$$

There are 53 nickels and 47 quarters. ∎

CAUTION

Notice that the equation comes from the total value column. Any equation must have like quantities on each side. You can never equate pounds with value, number of coins with the dollar value of the coins, etc.

Example 3

A chemist needs a 40% alcohol solution. How much pure alcohol must be added to nine quarts of water to obtain a mixture of 40% alcohol?

Solution

Here we are adding pure alcohol to 9 quarts of water.
Let x = the number of quarts of pure alcohol to be added.
Then $x + 9$ = the total number of quarts in the resulting mixture.

Type	Quarts ·	% Alcohol =	Amount of Alcohol
Alcohol	x	100%	$1x$
Mixture	$x + 9$	40%	$0.40(x + 9)$

Notice that the amount of alcohol in the mixture will be equal to the amount added.

$$\overbrace{x}^{\text{Amount of alcohol added}} = \overbrace{0.40(x+9)}^{\text{Amount of alcohol in the mixture}}$$

$$x = 0.40x + 3.6$$

$$0.60x = 3.6$$

$$x = 6$$

Therefore, six quarts of alcohol must be added. ∎

Example 4   Mary has $4.25 in nickels, dimes, and quarters. She has twice as many dimes as quarters. If the total number of coins is 37, find the number of each kind of coin.

Solution   Let q = the number of quarters.
Then $2q$ = the number of dimes.
Since Mary has 37 coins, we can subtract the number of quarters and dimes from 37 to get an expression for the number of nickels. Thus
$$37 - (q + 2q) = 37 - 3q = \text{the number of nickels.}$$

Type	Number	· Value	= Total Value
Quarters	q	0.25	$0.25q$
Dimes	$2q$	0.10	$0.10(2q)$
Nickels	$37 - 3q$	0.05	$0.05(37 - 3q)$
Mixture	37		4.25

The sum of the values of the coins is equal to the total value of the mixture of coins.

$$0.25q + 0.10(2q) + 0.05(37 - 3q) = 4.25$$

$$0.25q + 0.20q + 1.85 - 0.15q = 4.25$$

$$0.30q + 1.85 = 4.25$$

$$0.30q = 2.40$$

$$q = 8$$

Mary has 8 quarters, 16 dimes, and 13 nickels. ∎

Example 5   The capacity of a car's radiator is nine liters. The mixture of antifreeze and water is 30% antifreeze. How much of the mixture in the radiator must be drawn off and replaced with pure antifreeze to raise the percentage of the mixture to 65% antifreeze?

Solution   Keep in mind that the equation must equate like quantities. We can choose to make an equation equating antifreeze with antifreeze, or water with water, but *not* water with antifreeze. If we choose to equate antifreeze with antifreeze, we proceed as follows. Let x = the number of liters of the mixture to be drawn off.

	Liters	·	% Antifreeze	=	Liters of Antifreeze
Initial amt.	9		30%		0.30(9)
Drawn off	x		30%		0.30(x)
Added back	x		100%		1.00(x)
Final amt.	9		65%		0.65(9)

Note that the original amount of antifreeze, less the amount of antifreeze drained off, plus the amount of antifreeze added back will equal the final amount of antifreeze.

$$0.30(9) - 0.30(x) + 1.00(x) = 0.65(9)$$
$$2.7 + 0.7x = 5.85$$
$$0.7x = 3.15$$
$$x = 4.5$$

Replace 4.5 liters of the mixture with antifreeze. ∎

EXERCISE 3-4-1

Solve.

1. A candy dealer wishes to mix caramels that sell for $3.50 per pound with creams that sell for $2.75 per pound to obtain 30 pounds of a mixture that will sell for $3.00 per pound. How many pounds of each must be used?

Type	Pounds ·	Price	= Total Value
Caramels	x	3.50	
Creams	$30 - x$	2.75	
Mixture	30	3.00	

2. If peanuts sell for $2.50 per pound and cashews sell for $4.50 per pound, how many pounds of each must be used to produce 20 pounds of mixed nuts to sell for $3.25 per pound?

3. A total of 75 coins in nickels and quarters has a value of $15.15. How many coins of each type are there?

4. A total of 35 coins in dimes and quarters has a value of $5.60. How many of each type of coin are there?

5. A collection of coins has three times as many nickels as quarters. If the total value of the coins is $8.40, find the quantity of each type of coin.

6. How many liters of pure alcohol must be added to ten liters of water to obtain a mixture of 60% alcohol?

7. A dairy wishes to produce 225 gallons of a "low-fat" milk, which is 2% butterfat, by adding skim milk (no butterfat) to milk that is 3.6% butterfat. How much of each type of milk must the dairy use?

8. How much water should be added to five liters of a 25% salt solution to obtain a 15% salt solution?

Type	Liters	· % Salt	= Amt. of Salt
Water	x	0	
Original Solution	5	25%	
New Solution	$x + 5$	15%	

9. A nursery wishes to prepare 100 bags of a "weed and feed" mixture by mixing fertilizer and weed killer. The mixture is to sell for $7.80 per bag. How many bags of fertilizer that sells for $7.00 per bag and weed killer which sells for $9.00 per bag must be used?

10. How many liters of pure alcohol must be added to ten liters of 50% alcohol to obtain a mixture of 60% alcohol?

11. A nurse has twelve milliliters of a 16% solution and wishes to dilute it to a 6% solution. How much distilled water must be added to obtain the desired solution?

12. A 10% salt solution is added to a 25% salt solution to obtain ten liters of a 15% salt solution. How much of each is to be used?

13. A merchant mixes three types of tea— Oolong tea at $1.75 per ounce, Assam tea at $2.50 per ounce, and Keemun tea at $2.00 per ounce—to produce 50 ounces of a blend that will sell for $2.10 per ounce. If twice as much Oolong tea as Keemun tea is used, how much of each type is used?

14. A collection of 53 coins contains nickels, dimes, and quarters. The total value is $7.50. If there are seven more dimes than quarters, find the number of each type of coin.

15. A tank contains five liters of a 30% acid solution. How much must be drained off and replaced with a 50% acid solution to obtain five liters of a 38% acid solution?

CHAPTER 3 SUMMARY

The number in brackets refers to the section of the chapter that discusses that concept.

Terminology

- A **proportion** is a statement that two ratios are equal. [3–2]
- **Distance-rate-time problems** involve the formula $d = rt$, where r is an average or constant rate. [3–3]
- **Mixture problems** involve combining quantities of various concentrations. [3–4]

Rules and Procedures

Solving Word Problems

- To solve a word problem, follow these steps: [3–2]
  Understand the problem. Read the problem carefully and think about it.
  Devise a plan. Write variable expressions for the unknowns and an equation for the problem.
  Carry out the plan. Solve the equation.
  Look back. Be sure you have answered the question and your answer makes sense.

CHAPTER 3 REVIEW

Write an algebraic expression for each phrase.

1. The sum of a given number and 5

2. The value, in cents, of x quarters

3. The number of meters in y centimeters

Write the unknown quantities as variable expressions.

4. The sum of two numbers is 84. Write expressions for the numbers.

5. The length of a rectangle is seven inches more than three times the width. Write expressions for the width and length.

6. Write an expression for the simple interest earned when an amount, A, is invested for a year at 8%.

7. Write an expression for the total amount received at the end of one year if x dollars is invested at 10.2% simple interest.

Solve.

8. The length of one side of a triangle is twice that of the second side. The first side is also four centimeters less than that of the third side. If the perimeter is 64 cm, find the lengths of the three sides.

9. A meter stick is cut into two pieces so that one piece is six centimeters longer than the other. Find the lengths of the two pieces.

10. Mindy is seven years older than Carolyn. The sum of their ages is 63. Find their ages.

11. A backpack is on sale at a discount of 35%. The sale price is $31.20. Find the original price of the backpack.

12. A certain amount of money is invested for one year at 8% simple interest. The total principal and interest at the end of the year is $1620. Find the original amount invested.

13. Jordan is four years older than Julia, and Shasta is two years younger than twice Julia's age. How old is each if the sum of their ages is 58?

14. Florence has twice as many dimes as nickels, and two more quarters than dimes. The total value of the coins is $3.50. How many of each kind of coin does she have?

15. At the start of the semester there were twice as many students enrolled in psychology as in algebra, and three fewer students in history than in psychology. After the first week of classes, the algebra class had doubled in enrollment, four students had dropped psychology, and five had added history. If the total enrollment after the first week of classes was 148, how many students were enrolled in each class at the start of the semester?

16. A boat travels upstream and back in 12 hours. If the speed of the boat in still water is 25 kilometers per hour and the current in the river is 10 kilometers per hour, how far upstream did the boat travel?

17. How many grams of water must be evaporated from 90 grams of salt water that contains 5% salt to obtain a solution that is 9% salt?

18. Peoria and Fort Wayne are 435 kilometers apart. Linda leaves Fort Wayne and travels toward Peoria at a speed of 65 kilometers per hour. At the same time, John leaves Peoria and travels toward Fort Wayne. At what speed must John travel to meet Linda in three hours?

19. A 40-meter length of rope is cut into three pieces. The second piece is three times as long as the first, and the third piece is five meters shorter than twice the length of the second. Find the length of each piece.

20. A plane with a still air speed of 680 kilometers per hour makes an outgoing trip against a headwind in five hours and the return trip in three hours and thirty minutes. Find the speed of the headwind.

21. The length of a rectangle is four feet more than twice the width. If the width is tripled and the length is doubled, the perimeter of the new rectangle is twelve feet more than twice the perimeter of the original. Find the dimensions of the original rectangle.

22. A tank contains 30 liters of an 8% acid solution. How much pure acid should be added to the solution to obtain a 10% acid solution?

23. A boat leaves a dock and travels at the rate of 36 kilometers per hour. Twenty minutes later a second boat leaves the same dock and travels the same route at 50 kilometers per hour. How long will it take the second boat to be five kilometers behind the first boat?

24. A full two-liter bottle contains a 20% alcohol solution. How much of the solution must be poured out and replaced with pure alcohol to obtain two liters of a 30% alcohol solution?

25. Zack's snowmobile needs a ratio of oil to gas of 1 to 20. How many ounces of oil should he add to 2.5 gallons of gas? (Note: 1 gallon = 128 ounces)

26. Water is mixed with grape juice concentrate in a ratio of 3 to 1. How many ounces of water should be mixed with a 6-ounce can of grape juice concentrate? How many ounces of grape juice will this make?

CHAPTER 3 PRACTICE TEST

1. Write an algebraic expression for each.
   a. Six more than three times the width

1. a. _____

   b. Half of Justin's age, decreased by seven

   b. _____

   c. 9.4% of the retail price

   c. _____

   d. The number of weeks in d days

   d. _____

2. Erin has three more dimes than nickels and twice as many quarters as dimes. She has a total of $5.05. How many of each type of coin does Erin have?

2. _____

3. A piece of rope that is 63 meters long is cut into three pieces such that the second piece is twice as long as the first and the third piece is three meters longer than the second. Find the length of each piece.

3. _____

4. Tom can row at the rate of 12 kilometers per hour in still water. He rows upstream for two hours and returns in one hour and twelve minutes. Find the rate of the current.

4. _____

5. Duke's chain saw requires a ratio of oil to gas of 1 to 40. How many ounces of oil should he add to one-half gallon of gas?

5. _____

6. How many liters of water must be added to 10 liters of a 12% alcohol solution to obtain an 8% alcohol solution?

6. _____

Products and Factoring

Finding products of polynomials and factoring polynomials are two important skills in the study of algebra. In the first section of this chapter, we will discuss finding products of polynomials. In the remainder of the sections, we introduce you to the process of factoring polynomials.

4–1   THE PRODUCT OF POLYNOMIALS

OBJECTIVES

Upon completion of this section you should able to:
1. Multiply two polynomials.
2. Multiply two binomials using the shortcut method known as FOIL.
3. Recognize and use patterns that result in special products.

In this section we will study how to find the product of two polynomials. The distributive property can be used to find the product of two polynomials. The first polynomial factor can be thought of as a in $a(b + c) = ab + ac$. This is illustrated in the following examples.

Example 1   Expand: $(2x + 3)(x + 4)$

Solution   We can use the distributive property and think of the binomial $(2x + 3)$ as the factor a.

$$a \quad (b+c) = \quad a \quad b \quad + \quad a \quad c$$
$$(2x+3)(x+4) = (2x+3)x + (2x+3)4$$

We use the distributive property two more times to remove the parentheses.

$$2x^2 + 3x + 8x + 12$$

Note that each term of the first binomial is multiplied by each term of the second binomial.

Combine similar terms.

$$2x^2 + 11x + 12 \quad \blacksquare$$

This technique can be extended to multiply any two polynomials. In the next example we will multiply a binomial and a trinomial.

Example 2　　Expand: $(x+2)(x^2-3x+4)$

Solution　　　We apply the distributive property by multiplying each term of the trinomial by each term of the binomial.

$$(x+2)(x^2-3x+4)$$

$$x \cdot x^2 - x \cdot 3x + x \cdot 4 + 2 \cdot x^2 - 2 \cdot 3x + 2 \cdot 4$$

$$x^3 - 3x^2 + 4x + 2x^2 - 6x + 8$$

Combine similar terms.

$$x^3 - x^2 - 2x + 8 \quad \blacksquare$$

Example 3　　Expand: $(x-2)(x^2+2x+4)$

Solution　　　$x \cdot x^2 + x \cdot 2x + x \cdot 4 + (-2)x^2 + (-2)2x + (-2)4$

$$x^3 + 2x^2 + 4x - 2x^2 - 4x - 8$$

$$x^3 - 8 \quad \blacksquare$$

EXERCISE 4-1-1

Expand.

1. $(x+5)(x^2+3x+1)$ 　　　　　　**2.** $(2x-1)(x^2-4x+3)$

3. $(x-5)(x^2+5x+25)$ 　　　　　**4.** $(x+3)(x^2-3x+9)$

5. $(x-6)(2x^2-x+1)$ 　　　　　　**6.** $(x+1)(x^3+4x^2-x+3)$

7. $\left(5a+b\right)\left(3a^2+ab-4b^2\right)$

8. $\left(2a-b\right)\left(4a^2+2ab+b^2\right)$

9. $\left(x^2+xy+1\right)\left(x-xy+5\right)$

10. $\left(x+y+4\right)\left(x-y-2\right)$

Recall that Example 1 gave us $\left(2x+3\right)\left(x+4\right)=2x^2+11x+12$.   When we multiply each term of the first binomial by each term of the second binomial, we see the following pattern.

1. $(2x+3)\,(x+4)$ The product of the *first* terms in the binomials is $2x^2$, which is the first term of the answer.

2. $(2x+3)\,(x+4)$ The product of the *outer* terms in the binomials is $8x$.

3. $(2x+3)\,(x+4)$ The product of the *inner* terms in the binomials is $3x$.

4. $(2x+3)\,(x+4)$ The product of the *last* terms in the binomials is 12, which is the last term of the answer.

When the outer and inner products are similar, they combine to give the middle term of the answer. In this case, $8x+3x=11x$.

This method of multiplying two binomials is sometimes called the **FOIL** method. FOIL stands for the First, Outer, Inner, and Last terms in the binomials.

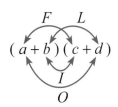

Example 4   Expand: $(3x + y)(2x - y)$

Solution

$$(3x + y)(2x - y)$$

$$6x^2 - 3xy + 2xy - y^2$$
$$6x^2 - xy - y^2 \quad \blacksquare$$

Example 5   Expand: $(2a + b)(3a + c)$

Solution $6a^2 + 2ac + 3ab + bc \quad \blacksquare$

The FOIL pattern is important since it is the basis for a factoring technique. It is important for you to be able to use this pattern.

EXERCISE 4-1-2
Expand.

1. $(a + 3)(a + 2)$ **2.** $(a + 1)(a + 5)$ **3.** $(x + 5)(x + 2)$

4. $(x - 1)(x - 4)$ **5.** $(a - 3)(a - 2)$ **6.** $(a - 4)(a - 2)$

7. $(a - 3)(a - 5)$ **8.** $(x + 2)(x - 3)$ **9.** $(x - 1)(x + 5)$

10. $(x + 7)(x - 2)$ **11.** $(a - 5)(a + 9)$ **12.** $(2x + 1)(x + 2)$

13. $(3x + 2)(2x + 1)$ **14.** $(a + 2)(3a - 5)$ **15.** $(2a - 1)(3a + 4)$

16. $(5x + 2)(x - 4)$ **17.** $(3x - 5)(2x - 3)$ **18.** $(2x + y)(3x + 5y)$

19. $(2a - b)(5a + 3)$ **20.** $(3x + 1)(x + 7y)$ **21.** $(a - 5)(4a + c)$

22. $(x - 2)(x - 2)$ **23.** $(m + 5)(m + 5)$ **24.** $(2n + 3)(2n + 3)$

We now will look at some special products. Consider the following examples.

Example 6   Expand: $(3x - 4)(3x + 4)$

Solution $9x^2 + 12x - 12x - 16$
$9x^2 - 16$ ∎

Example 7   Expand: $(2x + 1)(2x - 1)$

Solution $4x^2 - 2x + 2x - 1$
$4x^2 - 1$ ∎

Note that if two binomials are in the form $(a - b)(a + b)$, then they multiply to give a binomial product. The resulting product $a^2 - b^2$ is called a **difference of squares**.

RULE

$$(a - b)(a + b) = a^2 - b^2$$

Example 8   Expand: $(5x - 6)(5x + 6)$

Solution $25x^2 - 36$ ∎

   Next we will discuss the product that is the result of squaring a binomial. Remembering that an exponent tells us how many times to use the base as a factor, we proceed as follows.

Example 9   Expand: $(x-3)^2$

Solution $(x-3)(x-3)$

$x^2 - 3x - 3x + 9$

$x^2 - 6x + 9$ ∎

The resulting product when squaring a binomial is called a **perfect square trinomial**.

RULE

$$(a+b)^2 = (a+b)(a+b) \qquad \text{and} \qquad (a-b)^2 = (a-b)(a-b)$$
$$= a^2 + 2ab + b^2 \qquad\qquad\qquad = a^2 - 2ab + b^2$$

Example 10   Expand: $(2x+1)^2$

Solution $(2x+1)(2x+1)$

$4x^2 + 4x + 1$ ∎

EXERCISE 4-1-3
Expand.

1. $(x + 3)(x - 3)$ **2.** $(x + 5)(x - 5)$ **3.** $(2x - 7)(2x + 7)$

4. $(3x - 1)(3x + 1)$ **5.** $(x + y)(x - y)$ **6.** $(a - 7)(a + 7)$

7. $(x+y)^2$

8. $(x+4)^2$

9. $(x-6)^2$

10. $(2x+1)^2$

11. $(5x-2)^2$

12. $(4x-3)^2$

Practice your skills.
Expand.

13. $(x+3)(x^2-3x+9)$

14. $(a+1)^2$

15. $(2a+5)(3a-7)$

16. $(x-1)(x^2+x+1)$

17. $(2y+3)(2y-3)$

18. $(2x-y)(4x^2+2xy+y^2)$

19. $(2x-1)^2$

20. $(x-5)(7x-1)$

4–2   FACTORING POLYNOMIALS: THE GREATEST COMMON FACTOR

> ### OBJECTIVES
>
> Upon completion of this section you should be able to:
> 1. Determine which factors are common to all terms in a polynomial expression.
> 2. Write an expression as a product of a greatest common factor and a polynomial.
> 3. Factor expressions when the greatest common factor consists of more than one term.

In earlier chapters, the distinction between *terms* and *factors* has been stressed. You should remember that terms are added or subtracted and factors are multiplied.

> ### DEFINITION
>
> **Terms** occur in an indicated sum or difference. **Factors** occur in an indicated product.

> ### DEFINITION
>
> An expression is in **factored form** only if the *entire* expression is an indicated product.

Example 1   These expressions *are* in factored form.

$$2x(x+y)$$
$$(x+1)(3x-2)$$
$$(a+4)(a^2+3a-1) \quad \blacksquare$$

Example 2   These expressions are *not* in factored form.

$$x^2-4$$
$$2(x+y)+2z$$
$$(a+1)(a-1)-3 \quad \blacksquare$$

When determining if we have factored form, we must always consider the entire expression. Factors can be made up of terms and terms can contain factors, but *factored form* must conform to the definition.

> **DEFINITION**
>
> **Factoring** is a process of changing an expression from a sum or difference of terms to a product of factors.

Note that in this definition it is implied that the *value* of the expression is not changed – only its *form*.

The process of factoring is essential to the simplification of many algebraic expressions and is a useful tool in solving higher degree equations.

Factoring is a process that will "undo" multiplication. Given a polynomial, we want to find the factors that were multiplied to give the polynomial as a product. There are several techniques for factoring. Each technique corresponds to a method used to multiply polynomials.

The first technique for factoring that we will discuss finds the greatest common factor and uses the distributive property.

> **DEFINITION**
>
> The **greatest common factor (GCF)** of several terms is the greatest expression that is a factor of all the terms. It will consist of the largest number that is a factor of all the coefficients and each of the variables to the highest power common to all the terms.

To factor a polynomial expression by finding common factors, we can proceed as in the following example.

Example 3   Factor: $3x^2 + 6xy + 9xy^2$

Solution   First list the factors of each term.

$3x^2$ has factors $1, 3, x, x^2, \mathbf{3x}$, and $3x^2$.

$6xy$ has factors $1, 2, 3, 6, x, 2x, \mathbf{3x}, 6x, y, 2y, 3y, 6y, xy, 2xy, 3xy$, and $6xy$.

$9xy^2$ has factors $1, 3, 9, x, \mathbf{3x}, 9x, y, 3y, 9y, xy, 3xy, 9xy, y^2, 3y^2, 9y^2, xy^2, 3xy^2$ and $9xy^2$.

Next, look for factors that are common to all terms and find the greatest of these. This is the greatest common factor. In this case, the greatest common factor is $3x$.

Proceed by placing $3x$ before a set of parentheses.

$$3x\,(\qquad)$$

The terms within the parentheses are found by *dividing* each term of the original expression by $3x$.

$$3x\left(x + 2y + 3y^2\right)$$

The original expression is now changed to factored form. ∎

> **RULE**
>
> To factor a polynomial using the greatest common factor:
> 1. Find the GCF.
> 2. Write the indicated product of the GCF and the remaining factors of the terms.

To check your answer, keep in mind that factoring changes the *form* but not the value of an expression. If the answer in Example 3 is correct, it must be true that $3x\left(x+2y+3y^2\right)=3x^2+6xy+9xy^2$. Multiply to see that this is true. A second check is also necessary for factoring – we must be sure that the expression has been *completely factored*. In other words, "Did we find all common factors? Can we factor further?"

In Example 3, if we had only used the factor 3, the answer would be

$$3\left(x^2+2xy+3xy^2\right).$$

Multiplying to check, we find the answer is equal to the original expression. However, the factor x is still present in all terms of the polynomial factor. Hence the expression is not completely factored. From this point on, the directions to factor imply that you should factor *completely*.

> **DEFINITION**
>
> If a variable expression cannot be factored, it is said to be **prime**.

> **DEFINITION**
>
> For an expression to be **completely factored**, the answer must meet two criteria.
> 1. It must be possible to multiply the factored expression and get the original expression.
> 2. Each variable factor must be prime.

Example 4   Factor: $12x^3+6x^2+18x$

Solution   It should not be necessary to list the factors of each term. You should be able to mentally determine the greatest common factor. A good procedure to follow is to think of the coefficients and variables individually. In other words, don't attempt to find the complete GCF at once. Instead first get the number, then each letter involved. For this trinomial we see that 6 is the GCF of 12, 6, and 18, and that x is the GCF of the variable part of each term. Hence $12x^3+6x^2+18x$ can be factored as $6x\left(2x^2+x+3\right)$.

Multiplying, we get the original trinomial. We can also see that the terms within the parentheses have no other common factor, so we know the solution is correct. ∎

Example 5   Factor: $a^2b^2c^2 + 2ab^2c^2 - 3ab^3c^4$

Solution   The greatest common factor is ab^2c^2.

$$ab^2c^2\left(a + 2 - 3bc^2\right)\ \blacksquare$$

Check $ab^2c^2\left(a + 2 - 3bc^2\right)$
$a^2b^2c^2 + 2ab^2c^2 - 3ab^3c^4$

Example 6   Factor: $2x - 4y + 2$

Solution   The greatest common factor is 2.
$$2(x - 2y + 1)\ \blacksquare$$

Example 7   Factor: $5x^3y + 10x^2y^2 + 5xy^2$

Solution $$5xy\left(x^2 + 2xy + y\right)\ \blacksquare$$

Example 8   Factor: $3x^2y + 5x + 2y^2$

Solution   Since there is no common factor (except 1), this expression is *prime*. $\blacksquare$

EXERCISE 4-2-1

Factor. If prime, so state.

1. $12a + 10b$

2. $12x - 20y$

3. $a^2 + 5a$

4. $x^2 - xy$

5. $2a^2 + 6a$

6. $15x^2 - 35x$

7. $x^2 - x$

8. $6a^2 + 3a$

9. $8x^2y + 12xy$

10. $x^3y - x^2y^2$

11. $10a^2b - 2ab^2 + 6ab$

12. $15x^4y - 20x^2y^2 + 5x^2y$

13. $3x + 4y - 5z$

14. $6x + 21y - 27z$

15. $14a^2b - 35ab - 63a$

16. $12x^3y^2 + 20x^2y + 28x^2y^2$

17. $11a^3b^4 + 44ab^3 - 33a^2b^4$

18. $3a - 27b + 9c$

19. $4a^2 - 29b + 10ab$

20. $27x^3y^2 + 18x^2y^2 - 36x^2y^3$

A common factor does not need to be a monomial. For instance, in the expression $2y(x + 3) + 5(x + 3)$ we have two terms. They are $2y(x + 3)$ and $5(x + 3)$. In each of these terms we have a factor of $(x + 3)$ that is a binomial. This binomial factor, $(x + 3)$, is a common factor.

Example 9   Factor: $2y(x + 3) + 5(x + 3)$

Solution   The greatest common factor is $(x + 3)$.
$$(x + 3)\,(2y + 5)\ \blacksquare$$

Example 10   Factor: $3x^2(x + y) + 5x(x + y)$

Solution   The greatest common factor is $x(x + y)$.
$$x(x + y)(3x + 5)\ \blacksquare$$

Example 11   Factor: $5x(a + b) + (a + b)$

Solution $$(a + b)\,(5x + 1)\ \blacksquare$$

EXERCISE 4-2-2

Factor. If prime, so state.

1. $2x(x + 4) + 3(x + 4)$ **2.** $4a(a + b) + 3(a + b)$ **3.** $a(a - 1) + 5(a - 1)$

4. $2x(x - 5) + 7(x - 5)$ **5.** $6x(y + 4) - 7(y + 4)$ **6.** $3a(b - c) - 4(b - c)$

7. $3x(x - 2) + (x - 2)$ **8.** $10x(y + 1) - (y + 1)$ **9.** $7a(a - 4) - 15(a - 4)$

10. $14a(b + 1) + 5(b + 1)$ **11.** $x(x + 3) - (x + 3)$ **12.** $a(a + b) + b(a + b)$

13. $3x^2(x + 1) + 2x(x + 1)$ **14.** $a^2(a - 1) + 3a(a - 1)$ **15.** $4a(a + b) + a^2(a + b)$

16. $5a(a - b) - (a - b)$ **17.** $x(x - 1) - (x - 1)$ **18.** $6a^2(a + 1) + 9(a + 1)$

19. $8a^2(a + 4) + 2a(a + 4)$ **20.** $15x^3(x - 3) - 12x^2(x - 3)$

4–3   FACTORING BY GROUPING

> **OBJECTIVE**
> Upon completion of this section you should be able to factor by grouping.

An extension of the ideas presented in Section 4–2 applies to a method of factoring called **factoring by grouping**.

Example 1   Factor: $3ax + 6y + a^2x + 2ay$

Solution   First note that not all four terms in the expression have a common factor, but that some of them do. For instance, we can factor 3 from the first two terms, giving $3(ax + 2y)$. If we factor a from the remaining two terms, we get $a(ax + 2y)$.

$$3(ax + 2y) + a(ax + 2y)$$

This expression is not yet in factored form. However, we now have a common factor of $(ax + 2y)$.

$$(ax + 2y)\,(3 + a)$$

Check   Multiplying $(ax + 2y)\,(3 + a)$, we get the original expression

$3ax + 6y + a^2x + 2ay.$ ∎

This is an example of *factoring by grouping* since we "grouped" the terms two at a time. Notice that the technique requires two steps and three applications of the distributive property.

Example 2   Factor: $ax - ay + 2x - 2y$

Solution
$$a(x - y) + 2(x - y)$$
$$(x - y)(a + 2)$$ ∎

Example 3   Factor: $2ax + 3a + 4x + 6$

Solution
$$a(2x + 3) + 2(2x + 3)$$
$$(2x + 3)(a + 2)$$ ∎

The terms of a polynomial may not be arranged so that factoring by grouping can be accomplished. If this is the case, try rearranging the terms.

Example 4   Factor: $3ax + 2y + 3ay + 2x$

Solution   The first two terms have no common factor. However, the first and third terms have a GCF of $3a$, and the second and fourth terms have a GCF of 2. So we rearrange the terms to place the third term after the first and then factor by grouping.

$$3ax + 3ay + 2x + 2y$$
$$3a(x+y) + 2(x+y)$$
$$(x+y)(3a+2) \blacksquare$$

In all cases it is important to be sure that the factors within parentheses are exactly alike. This may require factoring a negative number or letter.

Example 5   Factor: $ax - ay - 2x + 2y$

Solution   Note that when we factor a from the first two terms we get $a(x - y)$. Looking at the last two terms, we see that factoring $+2$ would give $2(-x + y)$, but factoring -2 gives $-2(x - y)$. We want the terms within parentheses to be $(x - y)$, so we factor -2.

$$a(x-y) - 2(x-y)$$
$$(x-y)(a-2) \blacksquare$$

EXERCISE 4-3-1

Factor.

1. $ax + ay + 3x + 3y$ **2.** $ax + ay + 4x + 4y$ **3.** $ab + ac - 2b - 2c$

4. $3a - 3b + a^2 - ab$ **5.** $2ax + ay + 6x + 3y$ **6.** $2ax + 3a + 4x + 6$

7. $ax - ay + 2x - 2y$

8. $ax + 2ay - 3x - 6y$

9. $5x + 10y + ax + 2ay$

10. $2ax + 2ay - 5x - 5y$

11. $3ax - 2y - 3ay + 2x$

12. $ax + y + ay + x$

13. $2ax - y + 2ay - x$

14. $xy + 3y + 2x + 6$

15. $6ax - y + 2ay - 3x$

16. $cx - 4y + cy - 4x$

17. $3x - 2y - 6 + xy$

18. $xy - 8 - 2x + 4y$

19. $5y - 2x - 10 + xy$

20. $3y - x - 3 + xy$

4–4   FACTORING TRINOMIALS

> **OBJECTIVES**
>
> Upon completion of this section you should be able to:
> 1. Factor a trinomial having a first term coefficient of 1.
> 2. Find the factors of any factorable trinomial.

In Section 4–1 we established a pattern for multiplying two binomials (FOIL). Since many of the resulting products were trinomials, we can use this pattern to develop a technique to factor trinomials. This technique of factoring is called **trial and error**.

Example 1   Factor: $x^2 + 10x + 24$

Solution   Since this is a trinomial and the terms have no common factors, we will use the multiplication pattern to factor. First write parentheses for the 2 binomial factors. We now wish to fill in the terms of the binomials so that the pattern will give the original trinomial when we multiply.

$$x^2 + 10x + 24 \qquad \text{Recall:}$$

$$(\quad)(\quad)$$

The *first* term in the trinomial is x^2, so each of the first terms in the binomials must be an x.

$$(x\quad)(x\quad)$$

The *last* term in the trinomial is 24, so now we must find factors of 24 for the last terms of the binomials. Pairs of factors of 24 are: 1 and 24, 2 and 12, 3 and 8, 4 and 6. We fill in the last terms in the binomials with these factors and check.

$$(x+1)(x+24) = x^2 + 25x + 24$$

$$(x+2)(x+12) = x^2 + 14x + 24$$

$$(x+3)(x+8) = x^2 + 11x + 24$$

$$(x+4)(x+6) = x^2 + 10x + 24$$

Only the last product has a middle term of $10x$, and therefore the correct factoring is

$$(x+4)(x+6).$$

Notice that the sum of the outer and inner products must be $10x$. So we needed to find factors of 24 with a sum of 10. ■

Example 2   Factor: $x^2 - 10x + 24$

Solution   This time we must find factors of 24 that have a sum of -10. The product of two negative numbers is positive, and the sum of two negative numbers is negative. So the correct factoring is

$$(x-4)(x-6). \quad ■$$

Example 3   Factor: $x^2 - 2x - 24$

Solution   Since the third term is -24, we realize that one of the signs of the last terms in the binomials must be positive and one must be negative. This is because the product of a positive number and a negative number is negative. We need to find factors of -24 with a sum of -2. The correct factoring is

$$(x-6)(x+4). \quad ■$$

The following points will help as you factor trinomials.

1. When the sign of the third term is positive, both signs in the factors must be alike – and they must be the same as the sign of the middle term.
2. When the sign of the last term is negative, the signs in the factors must be unlike.

Remember, a polynomial that does not factor is *prime*. If you have tried all the possible binomial factors and none give the original trinomial, the trinomial is prime.

EXERCISE 4-4-1

Factor. If the expression is prime, so state.

1. $x^2 + 5x + 6$ 2. $x^2 + 7x + 6$ 3. $x^2 + x - 6$ 4. $x^2 - x - 6$

5. $x^2 - 5x - 6$ 6. $x^2 + 4x - 21$ 7. $x^2 + 7x - 18$ 8. $a^2 + 11a + 18$

9. $y^2 - 8y + 12$ **10.** $m^2 + m - 12$ **11.** $p^2 + 34p - 35$ **12.** $z^2 + 13z + 36$

13. $x^2 + 15x + 21$ **14.** $x^2 - 11x + 6$ **15.** $a^2 - 12ab + 35b^2$ **16.** $x^2 + 5xy - 36y^2$

17. $x^2 + 13x - 48$ **18.** $x^2 + 6x - 40$ **19.** $x^2 - 13x + 40$ **20.** $x^2 - 13x + 30$

21. $x^2 + 15x + 50$ **22.** $x^2 + 55x + 54$ **23.** $x^2 + 49x - 50$ **24.** $x^2 + 10x - 56$

25. $t^2 + 22t + 72$ **26.** $c^2 + 21cd + 20d^2$ **27.** $x^2 - 5xy - 24y^2$ **28.** $x^2 + 20x - 96$

29. $x^2 - 28x + 96$ **30.** $a^2 + 11a + 22$ **31.** $x^2 + 31x + 84$ **32.** $x^2 - 8x - 84$

33. $x^2 - 83x - 84$ **34.** $x^2 - 11x + 48$ **35.** $x^2 - 39x + 140$ **36.** $m^2 + 21mn - 22n^2$

37. $x^6 - x^3 - 20$ **38.** $a^4 + 16a^2 + 15$ **39.** $y^8 + y^4 - 30$ **40.** $z^6 - z^3 - 12$

In the previous exercise the coefficient of every first term was 1. When the coefficient of the first term is not 1, we must also consider factors of this coefficient. This increases the number of possible factors.

Example 4   Factor: $6x^2 + 17x + 12$

Solution   The pairs of factors of $6x^2$ we need to consider are: x and $6x$, $2x$ and $3x$. Pairs of factors of 12 are: 1 and 12, 2 and 6, 3 and 4.

Note that there are twelve ways to obtain the first and last terms, but only one has $17x$ as a middle term.

$(6x+1)(x+12)$	Middle term $73x$
$(6x+2)(x+6)$	Middle term $38x$
$(6x+3)(x+4)$	Middle term $27x$
$(6x+12)(x+1)$	Middle term $18x$
$(6x+6)(x+2)$	Middle term $18x$
$(6x+4)(x+3)$	Middle term $22x$
$(2x+2)(3x+6)$	Middle term $18x$
$(2x+6)(3x+2)$	Middle term $22x$
$(2x+1)(3x+12)$	Middle term $27x$
$(2x+12)(3x+1)$	Middle term $38x$
$(2x+4)(3x+3)$	Middle term $18x$
$(2x+3)(3x+4)$	Middle term $17x$

There is only one way to obtain all three terms:

$(2x+3)(3x+4)$. ■

In the preceding example one out of twelve possibilities is correct. Thus *trial and error* can be very time consuming unless we proceed systematically and employ our knowledge of addition and multiplication facts. In example 4 we would immediately eliminate many of the combinations. Since we are searching for $17x$ as a middle term we would not attempt those possibilities that multiply 6 by 6, or 3 by 12, or 6 by 12, and so on, as those products will be larger than 17. Also, since 17 is odd, we know it is the sum of an even number and an odd number. All of these things help reduce the number of possibilities to try.

> ## A PROCEDURE FOR TRIAL AND ERROR FACTORING
> 1. Find factors of the first and last terms of the trinomial.
> 2. Determine the signs in the binomial factors.
> 3. Eliminate impossible factor combinations.
> 4. Try reasonable combinations by testing outer + inner.

Example 5   Factor: $4x^2 - 5x - 6$

Solution   First we should analyze the problem.

1. The pairs of factors of $4x^2$ we need to consider are: x and $4x$, $2x$ and $2x$. The factors of 6 are: 1 and 6, 2 and 3.
2. The last term is negative, so use unlike signs.
3. We eliminate a product of $4x$ and 6 as too large.
4. Try combinations.

$(2x-3)(2x+2)$ This is incorrect because the second binomial factor has a factor of 2.

$(4x-2)(x+3)$ This is incorrect because the first binomial factor has a factor of 2.

$(4x-3)(x+2)$ Here the middle term is $+5x$, which is the right number but the wrong sign. Be careful not to accept this as the solution.

$(4x+3)(x-2)$ This gives the correct sign for the middle term. ■

Notice that we eliminated some of the possibilities in Example 5 because one of the binomial factors was not factored. If the terms of the original trinomial have no common factors, then the binomial factors must be prime.

Example 6   Factor: $6x^2 - 23x + 15$

Solution   Analyze.

1. The pairs of factors of $6x^2$ we need to consider are: x and $6x$, $2x$ and $3x$. The factors of 15 are: 1 and 15, 3 and 5.
2. The last term is positive, so use like signs. The middle term is negative, so both signs will be negative.
3. Eliminate as too large the product of 15 with $2x$, $3x$, or $6x$.
4. Try combinations.

$(2x-5)(3x-3)$ Not an option because 3 is a factor of $(3x-3)$

$(2x-3)(3x-5)$ Middle term $(-19x)$ INCORRECT

$(6x-3)(x-5)$ Not an option because 3 is a factor of $(6x-3)$

$(6x-5)(x-3)$ Middle term $(-23x)$ CORRECT ■

EXERCISE 4-4-2

Factor. If the expression is prime, so state.

1. $2x^2 + 5x + 2$ **2.** $2x^2 + 11x + 5$ **3.** $2x^2 + 5x + 3$ **4.** $3a^2 + 8a + 4$

5. $10y^2 + 19y + 6$ **6.** $3t^2 - 13t - 10$ **7.** $6c^2 + 7c + 2$ **8.** $6x^2 - x - 15$

9. $4x^2 - 11x + 6$ **10.** $4x^2 + 4x + 1$ **11.** $32x^2 + 4x - 3$ **12.** $15x^2 + 23x + 4$

13. $8x^2 - x - 3$ **14.** $12x^2 + x - 5$ **15.** $6x^2 + 31x + 5$ **16.** $10t^2 + 19t + 9$

17. $5a^2 + 8a + 3$ **18.** $12x^2 - 11x + 2$ **19.** $16x^2 - 6xy - y^2$ **20.** $9x^2 + 9xy + 2y^2$

21. $4a^2 - ab - 18b^2$ **22.** $15x^4 + 16x^2 - 28$ **23.** $15x^4 - 26x^2 - 24$ **24.** $8x^6 - 14x^3 - 15$

4–5 FACTORING TRINOMIALS – THE KEY NUMBER (optional)

OBJECTIVES

Upon completion of this section you should be able to:
1. Find the key number of a trinomial.
2. Use the key number to factor a trinomial.

Now we will discuss an alternative to trial and error factoring. This alternative may not always be practical for large numbers, but it may increase speed and accuracy for those who master it. To understand this technique, you must have studied factoring by grouping in Section 4–3.

The first step in this technique is finding the key number.

DEFINITION

When we factor trinomials, the **key number** is the product of the coefficients of the first and third terms.

Example 1   In $6x^2 - 22x + 12$, the key number is $(6)(12) = 72$. ∎

Example 2   In $5x^2 + 13x - 6$, the key number is $(5)(-6) = -30$. ∎

Example 3   Factor: $4x^2 + 3x - 10$

Solution   **Step 1** Find the key number $(4)(-10) = -40$.
            Step 2 Find factors of (-40) that will add to give the coefficient of the middle term $(+3)$. They are $(+8)$ and (-5) since $(+8) + (-5) = +3$.

            Step 3 Rewrite the original problem by breaking the middle term into the two parts found in step 2, $8x - 5x = 3x$.

$$4x^2 + 8x - 5x - 10.$$

            Step 4 Factor the result from step 3 by the grouping method studied in Section 4–3.

$$4x(x+2) - 5(x+2)$$
$$(x+2)(4x-5) \quad ∎$$

Example 4   Factor: $3x^2 - 10x - 8$

Solution   Key number = -24.

Factors of the key number that add to give -10 are -12 and 2.

$$3x^2 - 12x + 2x - 8$$
$$3x(x-4) + 2(x-4)$$
$$(x-4)(3x+2) \quad \blacksquare$$

The *key number* technique is a good alternative for two reasons.

1. You are able to make a more direct search for factors.
2. If no factors of the key number will combine to obtain the coefficient of the middle term, the trinomial cannot be factored.

EXERCISE 4-5-1

Use the key number to factor.

1. $x^2 + 9x + 20$ **2.** $x^2 - x - 6$ **3.** $x^2 - 8x + 12$ **4.** $2x^2 + 7x + 3$

5. $3x^2 + 5x + 2$ **6.** $2x^2 - 5x - 12$ **7.** $5x^2 + 9x - 2$ **8.** $2x^2 - 11x + 15$

9. $3x^2 - 14x + 8$ **10.** $3x^2 - 28x + 32$ **11.** $4x^2 + 9x - 9$ **12.** $x^2 + 8x + 16$

13. $4x^2 + 8x + 3$ **14.** $6x^2 - x - 12$ **15.** $16x^2 - 8x - 3$

4–6   FORMULAS FOR FACTORING SPECIAL PRODUCTS

OBJECTIVES

Upon completion of this section you should be able to:
1. Identify and factor the difference of two squares.
2. Identify and factor the sum or difference of two cubes.
3. Identify and factor a perfect square trinomial.

In this section we wish to examine some special cases of factoring. If these special cases are recognized, the factoring is then greatly simplified.

The first special case we will discuss is the **difference of squares**. Recall from Section 4–1 that $(a+b)(a-b)=a^2-b^2$. Reading this equation from right to left, we have a rule for factoring a difference of squares.

RULE

$$a^2-b^2=(a+b)(a-b)$$

Example 1   Factor: $25x^2-16$

Solution   Here both terms are squares and they are separated by a negative sign.
The form of this problem is $(5x)^2-(4)^2$, which is the form a^2-b^2.

$$(5x+4)(5x-4) \quad \blacksquare$$

Special cases do make factoring easier, but be certain to recognize that a special case is just that-very special. In this case *both* terms must be squares *and* the sign must be negative, hence "the difference of squares." You must be able to recognize square numbers. A list of square numbers is included on the inside back cover of the text. Also, if the exponents on the variables are even, the variable factor of the term is a square.

Example 2 x^2-7 This is not the special case because 7 is not a square. $\blacksquare$

Example 3 $4x^2+9$ This is not the special case because it is a sum. $\blacksquare$

CAUTION   Students often overlook the fact that 1 *is* a perfect square. Thus an expression such as x^2-1 is a difference of squares and can be factored by this method.

Example 4   Factor: $x^2 - 1$

Solution $(x+1)(x-1)$ ▪

Example 5   Factor: $4x^2 - 9$

Solution $(2x+3)(2x-3)$ ▪

Example 6   Factor: $x^4 - 25$

Solution $(x^2 - 5)(x^2 + 5)$ ▪

Example 7   Factor: $(a+b)^2 - c^2$

Solution $\left[(a+b)+c\right]\left[(a+b)-c\right] = (a+b+c)(a+b-c)$ ▪

EXERCISE 4-6-1

Factor.

1. $x^2 - 4$ **2.** $x^2 - 1$ **3.** $a^2 - 81$ **4.** $x^2 - 100$

5. $y^2 - 121$ **6.** $4x^2 - 9$ **7.** $16x^2 - 1$ **8.** $4x^2 - y^2$

9. $4x^2 - 25y^2$ **10.** $1 - x^2$ **11.** $4 - x^2$ **12.** $9 - 4x^2$

13. $16a^2 - 121b^2$ **14.** $(x+y)^2 - 64$ **15.** $x^4 - 9$ **16.** $x^6 - 36$

17. $25a^4 - 49$ **18.** $(x+y)^4 - z^2$

There are two more special binomials that can be factored. If we expand factors that fit the pattern

$$(a+b)(a^2 - ab + b^2), \text{ we get}$$

$$a^3 - a^2b + ab^2 + a^2b - ab^2 + b^3, \text{ and finally}$$

$$a^3 + b^3, \text{ which is a product known as a } \textbf{sum of cubes}.$$

Likewise, expanding

$$(a-b)(a^2 + ab + b^2) \text{ as}$$

$$a^3 + a^2b + ab^2 - a^2b - ab^2 - b^3 \text{ gives}$$

$$a^3 - b^3, \text{ which is a } \textbf{difference of cubes}.$$

Therefore, we can factor sums and differences of cubes using the following rule.

RULE

$$a^3 + b^3 = (a+b)(a^2 - ab + b^2)$$

$$a^3 - b^3 = (a-b)(a^2 + ab + b^2)$$

To use this rule successfully, you must be able to recognize expressions that are cubes. For a list of cubes, see the back cover of the text.

Example 8   Factor: $x^3 + 27$

Solution   We must first recognize that this is the sum of cubes.

$$x^3 + 27$$

$$x^3 + (3)^3$$

Now follow the pattern in the rule. Let x take the place of a and 3 take the place of b.

$$(x+3)\left[x^2 - 3x + (3)^2 \right]$$

$$(x+3)(x^2 - 3x + 9) \quad \blacksquare$$

Example 9   Factor: $8x^3 - 125$

Solution

$$(2x)^3 - (5)^3$$

$$(2x-5)\left[(2x)^2 + (2x)(5) + (5)^2 \right]$$

$$(2x-5)(4x^2 + 10x + 25) \quad \blacksquare$$

EXERCISE 4-6-2

Factor.

1. $x^3 + y^3$

2. $x^3 - y^3$

3. $a^3 - 8$

4. $x^3 + 8$

5. $x^3 - 1$

6. $a^3 + 1$

7. $27a^3 + 1$

8. $8x^3 - y^3$

9. $27a^3 + 8b^3$

10. $a^3 - 125$

11. $64x^3 - 27y^3$

12. $x^3 - y^6$

A third special case in factoring is the **perfect square trinomial**. In Section 4–1, we saw that when we square a binomial, the product is a perfect square trinomial. So a perfect square trinomial will factor as a binomial squared.

RULE

$$a^2 + 2ab + b^2 = (a+b)^2$$

$$a^2 - 2ab + b^2 = (a-b)^2$$

We recognize this case by noting three special features.

1. The first term is a square.
2. The third term is a square.
3. The middle term is twice the product of the square roots of the first and third terms.

squares

$$\boxed{a^2} + \boxed{2ab} + \boxed{b^2}$$

twice the product of the square roots of the first and third terms

Example 10   Factor: $25x^2 + 20x + 4$

Solution   **1.** $25x^2$ is a square since $(5x)^2 = (5x)(5x) = 25x^2$.

2. 4 is a square since $2^2 = (2)(2) = 4$.

3. $20x$ is twice the product of the square roots of $25x^2$ and 4. $20x = 2(5x)(2)$

$$25x^2 + 20x + 4 = (5x + 2)^2 \quad \blacksquare$$

Example 11   Factor: $25x^2 - 20x + 4$

Solution $(5x - 2)^2 \quad \blacksquare$

Example 12   Factor: $9x^2 - 6x + 1$

Solution $(3x - 1)^2 \quad \blacksquare$

Example 13   Factor: $x^2 + 8x + 16$

Solution $(x + 4)^2 \quad \blacksquare$

Example 14   Factor: $4x^2 + 13x + 9$

Solution   This is not the special case of a perfect square trinomial. The middle term would need to be $12x$ to have a perfect square trinomial. However, this trinomial can be factored.

$$(4x + 9)(x + 1) \quad \blacksquare$$

EXERCISE 4-6-3

Factor.

1. $x^2 + 6x + 9$ **2.** $x^2 + 10x + 25$ **3.** $x^2 - 14x + 49$

4. $a^2 - 2a + 1$ **5.** $4x^2 + 12x + 9$ **6.** $4a^2 - 36ab + 81b^2$

7. $9x^2 + 30x + 25$ **8.** $49a^2 + 14a + 1$ **9.** $36x^2 - 60xy + 25y^2$

4–7   COMPLETE FACTORIZATION

> **OBJECTIVE**
>
> Upon completion of this section you should be able to completely factor a polynomial of 2, 3, or 4 terms by applying the methods of this chapter.

We have now studied all of the usual techniques of factoring found in elementary algebra. However, you must be aware that a single problem can require more than one of these methods. Remember that there are two checks necessary for correct factoring.

1. Will the factors multiply to give the original problem?
2. Are all variable factors prime?

We suggest the following factoring strategy.

> **STRATEGY FOR COMPLETELY FACTORING A POLYNOMIAL**
>
> 1. Factor the greatest common factor (GCF) of all the terms in the polynomial.
>
> 2. Try to factor a binomial as
>
>   **a.** the difference of two squares,
> $$a^2 - b^2 = (a+b)(a-b),$$
>   or
>
>   **b.** the sum or difference of two cubes,
> $$a^3 + b^3 = (a+b)(a^2 - ab + b^2)$$
> $$a^3 - b^3 = (a-b)(a^2 + ab + b^2).$$
>
> 3. Try to factor a trinomial
>
>   **a.** as a perfect square trinomial,
> $$a^2 + 2ab + b^2 = (a+b)^2$$
> $$a^2 - 2ab + b^2 = (a-b)^2,$$
>   or
>
>   **b.** using trial and error (reverse FOIL).
>
> 4. Try to factor larger polynomials by grouping.
>
> 5. After each factorization, examine the new factors to see whether any can be factored further.

Example 1   Factor: $2x^2 + 10x + 12$

Solution   We notice that 2 is a common factor.

$$2(x^2 + 5x + 6)$$

Now if we leave the problem at this point, the first check works but the second doesn't.

We see that the factor $(x^2 + 5x + 6)$ can now be factored into

$$(x+3)(x+2).$$

The complete factorization is

$$2(x+3)(x+2). \quad \blacksquare$$

Suppose in this example we had looked at the problem $2x^2 + 10x + 12$ and simply used the trial and error method. We could get

$$(2x+4)(x+3).$$

Once again the first check works because the product gives the original trinomial. However, the factor $(2x+4)$ has a common factor of 2.

$$2x + 4 = 2(x+2)$$

Thus $2x^2 + 10x + 12 = 2(x+2)(x+3).$

This factorization is complete because it meets both criteria. When factoring, remember to *factor completely*.

Example 2   Factor: $12x^2 - 27$

Solution   We note a common factor of 3.

$$3(4x^2 - 9)$$

At this point we recognize $4x^2 - 9$ as the difference of squares.

$$3(2x-3)(2x+3) \quad \blacksquare$$

Example 3   Factor: $16ax^4 - 2ax$

Solution $2ax(8x^3 - 1)$

$2ax(2x-1)(4x^2 + 2x + 1) \quad \blacksquare$

EXERCISE 4-7-1

Factor. If the expression is prime, so state.

1. $4x^2 + 14x + 6$

2. $6x^2 + 5x + 1$

3. $3x^2 + 6x - 144$

4. $3x^2 - 8x - 3$

5. $3x^2 - 12$

6. $4x^2 + 24x + 36$

7. $x^2 + x + 1$

8. $3a^3 + 10a^2 - 8a$

9. $42b^2 - 7b - 7$

10. $5x^2 + 5x - 60$

11. $x^3 - 8x^2 + 12x$

12. $20x^2 - 5$

13. $ax^2 - 3 + 3x^2 - a$

14. $14x^2 + 11x + 2$

15. $8x^2 - 8x + 2$

16. $3t^3 - 39t^2 + 120t$

17. $m^2 + 5m - 14$

18. $6x^2 - 3x + 2$

19. $4x^2 - 4x - 3$

20. $a^2x + a^2 - 9x - 9$

21. $27x^3 - 75x$

22. $2z^2 + 4z + 6$

23. $x^2 + x - 30$

24. $3x^2 + 93x + 252$

25. $7x^2 + 3x - 2$

26. $4x^4 - 32x$

27. $5x^3 - 5$

4–8   LONG DIVISION OF POLYNOMIALS

> **OBJECTIVES**
>
> Upon completion of this section you should be able to:
> 1. Divide a polynomial by a binomial.
> 2. Determine if a binomial is an exact divisor (factor) of a polynomial.

From our discussion of the relationship between the factored form and the expanded form of a polynomial, it should be clear to you that any factor will divide into the original expanded form without a remainder. Another name for a factor is an **exact divisor**. Since $(x + 1)$ is a factor of $x^2 - 3x - 4$, it is also an exact divisor.

The factoring techniques we have presented in this chapter do not give us the tools to factor all polynomials. However, we can use a technique called **long division** to help us determine if a given binomial is a factor of a polynomial, and then it may be possible to continue the factoring process.

Example 1   Is $(x + 3)$ an exact divisor (factor) of $2x^3 + 11x^2 + 8x - 21$?

Solution   To answer this question, we proceed as follows. This is the long division technique.

Step 1   Be sure that both polynomials are in descending powers of the variable and supply a zero for any missing terms. Then arrange as follows:

$$x+3 \overline{)2x^3 + 11x^2 + 8x - 21}$$

The divisor is $(x + 3)$, the dividend is $2x^3 + 11x^2 + 8x - 21$, and the result is called the quotient.

Step 2   To obtain the first term of the quotient, we must find an expression which, when multiplied by the first term in the divisor, x, will give the first term in the dividend, $2x^3$. The result, $2x^2$, should be placed directly above the $11x^2$ term in the dividend. It is good practice to keep all like terms aligned. We record this as follows:

$$\begin{array}{r} 2x^2 \\ x+3 \overline{)2x^3 + 11x^2 + 8x - 21} \end{array}$$

Step 3   Multiply the *entire* divisor by the term obtained in step 2, $2x^2$, and subtract the result from the dividend. After subtracting, bring down the next term, $8x$. Then align like terms as follows:

$$\begin{array}{r} 2x^2 \\ x+3 \overline{)2x^3 + 11x^2 + 8x - 21} \\ \underline{2x^3 + 6x^2 } \\ 5x^2 + 8x \end{array}$$

Step 4   To obtain the next term of the quotient, we must find an expression which, when multiplied by the first term in the divisor, x, will give $5x^2$. Place the result, $5x$, directly above the $8x$ term in the dividend. Multiply, align like terms, and subtract to get $-7x$. We record this as follows:

$$
\begin{array}{r}
2x^2 +\ 5x \\
x+3\overline{\smash{)}2x^3 +11x^2 +\ 8x\ -\ 21} \\
\underline{2x^3 +\ 6x^2 } \\
5x^2 +\ 8x \\
\underline{5x^2 + 15x } \\
-\ 7x
\end{array}
$$

This process is repeated until either the remainder is zero or the power of the first term of the remainder is less than the power of the first term of the divisor.

$$
\begin{array}{r}
2x^2 +\ 5x\ -\ 7 \\
x+3\overline{\smash{)}2x^3 +11x^2 +\ 8x\ -\ 21} \\
\underline{2x^3 +\ 6x^2 } \\
5x^2 +\ 8x\ -\ 21 \\
\underline{5x^2 + 15x } \\
-\ 7x\ -\ 21 \\
\underline{-\ 7x\ -\ 21} \\
0
\end{array}
$$

Since the remainder is zero, we know that $(x + 3)$ is an exact divisor (factor) of $2x^3 +11x^2 +8x-21$. In fact, we know that

$$(x+3)(2x^2+5x-7) = 2x^3 +11x^2 +8x-21.$$

We can use this information to factor the expression $2x^3 +11x^2 +8x-21$ as

$$(x+3)(2x^2+5x-7)$$

and finally as

$$(x+3)(2x+7)(x-1). \qquad \blacksquare$$

Example 2   Is $5x + 1$ a factor of $5x^3 - 24x^2 + 7$?

Solution   Notice that the x term is missing in the dividend. This means that the coefficient of that term is zero and we must enter it as a term as follows:

$$
\begin{array}{r}
x^2 \;-\; 5x \;+\; 1 \\
5x+1\overline{\smash{\big)}\,5x^3 \;-\; 24x^2 \;+\; 0x \;+\; 7} \\
\underline{5x^3 \;+\;x^2} \\
-\;25x^2 \;+\; 0x \\
\underline{-\;25x^2 \;-\; 5x} \\
5x \;+\; 7 \\
\underline{5x \;+\; 1} \\
6
\end{array}
$$

Since the remainder is not zero, we know that $5x + 1$ is *not* a factor of $5x^3 - 24x^2 + 7$. ∎

EXERCISE 4-8-1

Find the quotient and remainder. Then answer the question.

1. Is $x + 7$ a factor of $x^2 + 5x - 14$?

2. Is $x + 2$ a factor of $x^3 - x^2 - 2x + 2$?

3. Is $x - 3$ an exact divisor of $2x^3 - 2x^2 - 5x + 6$?

4. Is $x - 5$ an exact divisor of $x^3 + 2x^2 - 30x + 2$?

5. Is $x + 4$ an exact divisor of $x^3 + 2x^2 + 32$?

6. Is $x + 3$ an exact divisor of $x^4 + x^2 - 90$?

7. Is $x - 10$ a factor of $x^3 + x^2 + 10x + 100$?

8. Is $3x + 2$ a factor of $3x^3 + 11x^2 - 9x - 10$?

9. Factor $4x^3 + 8x^2 - x - 2$ given that $2x + 1$ is a factor.

10. Factor $6x^3 - 18x + 12$ given that $2x + 4$ is a factor.

CHAPTER 4 SUMMARY

The number in brackets refers to the section of the chapter that discusses that concept.

Terminology

- The **FOIL** method is a pattern used to multiply two binomials. [4–1]
- A **difference of squares** is a binomial of the form $a^2 - b^2$. [4–1]
- A **perfect square trinomial** is a trinomial that is the result of squaring a binomial. [4–1]
- An expression is in **factored form** only if the entire expression is an indicated product. [4–2]
- **Factoring** is a process of changing an expression from a sum or difference of terms to a product of factors. [4–2]
- The **greatest common factor** is the largest factor common to all terms. [4–2]
- A **prime** expression cannot be factored. [4–2]
- An expression is **completely factored** when no further factoring is possible. [4–2]
- The possibility of **factoring by grouping** exists when an expression contains four or more terms. [4–3]
- The **key number** is the product of the coefficients of the first and third terms of a trinomial. [4–5]
- A **sum of cubes** is a binomial of the form $a^3 + b^3$. [4–6]
- A **difference of cubes** is a binomial of the form $a^3 - b^3$. [4–6]
- An **exact divisor** of a polynomial is a factor of the polynomial. [4–8]

Rules and Procedures

- The distributive property can be used to find the product of any two polynomials. [4–1]
- To multiply two binomials use the FOIL method. [4–1]

- For factoring to be correct the solution must meet two criteria:

    1. It must be possible to expand the factored expression and obtain the original expression.
    2. Each variable factor must be prime. [4–2]

- Factoring Strategy [4–7]
    1. Factor the GCF of all the terms of the polynomial. [4–2]
    2. Try to factor a binomial as
        a. the difference of squares,

        $$a^2 - b^2 = (a - b)(a + b), \text{ or}$$

        b. the sum or difference of cubes,

        $$a^3 + b^3 = (a + b)(a^2 - ab + b^2)$$
        $$a^3 - b^3 = (a - b)(a^2 + ab + b^2).$$

        [4–6]
    3. Try to factor a trinomial
        a. as a perfect square trinomial [4–6]

        $$a^2 + 2ab + b^2 = (a + b)^2$$
        $$a^2 - 2ab + b^2 = (a - b)^2, \text{ or}$$

        b. using trial and error. [4–4]

4. Try to factor larger polynomials by grouping. [4–3]

5. After each factorization, examine the new factors to see whether any can be factored further.

- If a polynomial is divided by a binomial and the remainder is zero, the binomial is an exact divisor of the polynomial. [4–8]

- Use **long division** to determine if a given binomial is a factor of a polynomial.

CHAPTER 4 REVIEW

Expand.

1. $(2x+3)(x-2)$

2. $(3a-2)^2$

3. $(3x+8)(3x-8)$

4. $(x+3)(2x^2-x+4)$

5. $(x+y+3)(x-y-1)$

Factor. If prime, so state.

6. $9x+6y$

7. a^2-a

8. $6x^3y+10x^2y^2$

9. $6a^2b+9ab^2-6ab$

10. $x(a+b)-2(a+b)$

11. $25x^2-49$

12. $x^2 + 8x + 16$

13. $9x^2 - 30x + 25$

14. $45 - 5x^2$

15. $24x^3 - 3$

16. $x^2 + 5x + 6$

17. $x^2 + 5x - 14$

18. $x^2 - 4x + 3$

19. $15 - 2x - x^2$

20. $72 - x - x^2$

21. $2x^2 + 7x + 2$

22. $2x^2 - x - 3$

23. $3a^2 + 13a - 10$

24. $6x^2 - 5x - 4$

25. $6t^2 - 13t + 2$

26. $4m^2 + 14m + 6$

27. $18x^2 + 3x - 6$

28. $6x^3 + 8x^2 - 8x$

29. $3x^4 + 9x^2 + 15$

30. $5z^6 + 13z^3 + 6$ **31.** $cx + dx + cy + dy$ **32.** $ax - 3y - 3x + ay$

33. $ax + 20 - 5a - 4x$ **34.** $a^2x - 8 + 2a^2 - 4x$ **35.** $2a^2x - 3 - 2x + 3a^2$

Find the quotient and remainder. Then answer the question.

36. Is $x - 1$ a factor of $x^3 + 2x^2 - x - 2$?   **37.** Is $x + 5$ an exact divisor of
$x^3 - 2x^2 - 20x + 75$?

38. Is $x - 3$ an exact divisor of $2x^3 - 4x^2 + 5x - 3$?   **39.** Is $3x - 1$ a factor of $3x^3 + 2x^2 - 4x + 1$?

40. Is $2x + 3$ a factor of $4x^3 + 4x^2 - x - 1$?

CHAPTER 4 PRACTICE TEST

Expand.

1. $(3x-1)(2x+3)$

1. _____

2. $(2x-1)(x^2-2x+3)$

2. _____

Factor. If prime, so state.

3. $22a^2+33a$

3. _____

4. $10x^3y-6x^2y^2+2xy^3$

4. _____

5. $2x(a-1)-3(a-1)$

5. _____

6. p^2-81

6. _____

7. $x^2-22x+121$

7. _____

8. $ax+bx+3a+3b$

8. _____

9. $x^2 - 14x + 49$

9. _____

10. $4a^2 - 19a + 12$

10. _____

11. $2x^2 + 13x + 15$

11. _____

12. $m^2 + 11m + 18$

12. _____

13. $25x^2 - 144$

13. _____

14. $r^2 - 6r - 27$

14. _____

15. $ax^2 - 27 + 3x^2 - 9a$

15. _____

16. $6x^2 - 10x - 4$

16. _____

17. $32x^3 - 72x$

17. _____

18. $z^2 + 2z - 63$ **18.** _____

19. $6y^3 + 3y^2 - 18y$ **19.** _____

20. $27x^3 + 36x^2 + 12x$ **20.** _____

21. $6t^2 - t - 15$ **21.** _____

22. $24x^2 + 52x + 20$ **22.** _____

Find the quotient and remainder. Then answer the question.

23. Is $x - 5$ an exact divisor of   **23.** _____
$3x^3 - 13x^2 - 8x - 10$?

24. Is $2x - 1$ a factor of $4x^3 + 4x^2 - 7x - 2$?   **24.** _____

MID BOOK TEST

1. Evaluate: $7 - (-3) + (-8)$ **1.** _____

2. Simplify: $\dfrac{-32x^7}{-4x}$ **2.** _____

3. Multiply: $5x^2y\left(3xy^3 - 2x^2\right)$ **3.** _____

4. Evaluate: $3^2 - 2\left[2^3 + 6 \div 3 \cdot 5\right]$ **4.** _____

5. Simplify: $10x^2 - \left[7x - 2x\left(5x + 2\right)\right]$ **5.** _____

6. Evaluate $6a^3 - 2a^2b + 3b^3$ when $a = -2$ and $b = 3$.   **6.** _____

Solve for x.

7. $5(x + 3) - 3 = 2(x - 3)$ **7.** _____

8. $\dfrac{x}{2} - 2 = \dfrac{x}{3}$

8. _____

9. $x - 7 = 1 - \dfrac{2x-1}{3}$

9. _____

10. $3(x-a) = \dfrac{1}{2}(x+3a)$

10. _____

Solve for x and graph the solution on the number line.

11. $-2x > 8$

11. _____

12. $|x-4| \le 5$

12. _____

Factor completely.

13. $14a^2b^3 - 7a^4b^2 + 21a^3b^3$

13. _____

14. $18y^2 - 8$ **14.** _____

15. $4m^2 - 20m + 25$ **15.** _____

16. $14x^2 + 21x - 14$ **16.** _____

17. $ax^2 - a - bx^2 + b$ **17.** _____

18. $27p^3 + 1$ **18.** _____

19. Is $x - 3$ a factor of $x^3 + 2x^2 - 17x + 6$?   **19.** _____
Show the division.

20. Write an algebraic expression for the number of hours in *m* minutes.

20. _____

21. Jena invested in stocks at an annual interest rate of 6%. She invested twice as much in bonds at an annual rate of 10%. If the total interest for the year on the two investments was $910, how much was invested at each rate?

21. _____

22. Melinda has 151 cm of ribbon. She needs to cut three pieces so that the first piece is 12 cm longer than the second and the third piece is 20 cm shorter than the first. How long should each piece be?

22. _____

23. A motorboat travels up a river against a 5 mph current for 2 hours and returns in $1\frac{1}{2}$ hours. Find the rate of the motorboat in still water.

23. _____

24. How many liters of pure alcohol must be added to 15 liters of water to obtain a solution of 40% alcohol?

24. _____

CHAPTER 5

SURVEY

The following questions refer to material discussed in this chapter. Work as many problems as you can and check your answers with the answer key in the back of the book. The results will direct you to the sections of the chapter in which you need to work. If you answer all questions correctly, you may already have a good understanding of the material contained in this chapter.

1. Simplify: $\dfrac{x^2-9}{x^2-x-6}$

    1. _____

2. Multiply: $\dfrac{x^2-8x+16}{2x^2-7x-4} \cdot \dfrac{4x^2-1}{x^2-x-12}$

    2. _____

3. Divide: $\dfrac{y^2-y-2}{y^2-5y+6} \div \dfrac{y^2+3y+2}{y^2+2y}$

    3. _____

4. Find the LCD for the following fractions:

    $\dfrac{1}{x^2+10x+24}, \dfrac{x-1}{x^2-16}$

    4. _____

5. Rewrite $\dfrac{x-3}{x+3}$ so that it has the denominator $(2x-7)(x+3)$.

    5. _____

6. Add: $\dfrac{5}{x^2+2x-8}+\dfrac{x+3}{x^2-3x+2}$

6. _____

7. Subtract: $\dfrac{x+5}{x^2+2x-3}-\dfrac{1}{x+3}$

7. _____

8. Simplify: $\dfrac{\dfrac{1}{x-3}-\dfrac{1}{x}}{3}$

8. _____

9. Solve: $\dfrac{2}{x}+\dfrac{1}{x-1}=\dfrac{4}{x}$

9. _____

10. Dawn can complete a job in 3 hours, and Erin can complete the same job in 6 hours. How long will it take both of them working together to complete the job?

10. _____

Rational Expressions

Fractions occur in algebra both in expressions and equations. Your knowledge of the arithmetic of fractions will form a foundation for work in rational expressions. A **rational expression** is the quotient of two polynomials. The methods used in simplifying, multiplying, dividing, adding, and subtracting rational expressions are identical with the methods from arithmetic for working with fractions (rational numbers). Equations with rational expressions are also covered in this chapter.

To work the problems in this chapter you must be proficient with the techniques of factoring from the previous chapter.

5–1   SIMPLIFYING RATIONAL EXPRESSIONS

OBJECTIVE
Upon completion of this section you should be able to simplify rational expressions.

The **fundamental principle of fractions** states that the numerator and denominator of a fraction can be multiplied by the same nonzero number without changing the value of the fraction.

RULE
$$\frac{a}{b} = \frac{ax}{bx}, \quad x \neq 0, b \neq 0$$

The fundamental principle is used to simplify fractions as well as to rewrite a fraction as an equivalent fraction with a different denominator and numerator. In this section, we will simplify rational expressions using the fundamental principle of fractions.

$$\frac{ax}{bx} = \frac{a}{b} \cdot \frac{x}{x} = \frac{a}{b} \cdot 1 = \frac{a}{b}, \quad x \neq 0, b \neq 0$$

In the above expression, the factor x is said to be common to the numerator and the denominator. When a common factor occurs, we will say that we are *canceling the common factor* and use a *strike through* to indicate that we actually have a factor of 1.

$$\frac{a\cancel{x}}{b\cancel{x}} = \frac{a}{b}, \quad x \neq 0, b \neq 0$$

A fraction is in **simplest form** if the numerator and denominator have no common factors.

> **RULE**
>
> To simplify a fraction or rational expression, factor both the numerator and denominator and then apply the fundamental principle of fractions (cancel common factors).

In studying rational expressions we will frequently use an example from arithmetic to illustrate a technique.

Example 1   Simplify: $\dfrac{12}{18}$

Solution

$$\frac{12}{18} = \frac{(2)(\cancel{6})}{(3)(\cancel{6})} = \frac{2}{3}$$

Notice that $\dfrac{12}{18}$ and $\dfrac{2}{3}$ have the same value but different forms. ■

Example 2   Simplify: $\dfrac{5x+20}{25}$

Solution   Factor. $\dfrac{5(x+4)}{25}$

Next, look for common factors and cancel. $\dfrac{\cancel{5}(x+4)}{(\cancel{5})(5)}$

The result is in simplest form. $\dfrac{x+4}{5}$ ■

Example 3   Simplify: $\dfrac{a^2 - b^2}{a^2 + 3ab + 2b^2}$

Solution

$$\frac{(a-b)(a+b)}{(a+b)(a+2b)}$$

$$\frac{(a-b)\cancel{(a+b)}}{\cancel{(a+b)}(a+2b)}$$

$$\frac{a-b}{a+2b} \quad ■$$

CAUTION   Note that the fundamental principle of fractions applies only to *factors*, not *terms*. For instance, $\dfrac{2+4}{2} = \dfrac{6}{2} = \dfrac{\cancel{2}\cdot 3}{\cancel{2}} = 3$ is true. However, $3 = \dfrac{6}{2} = \dfrac{\cancel{2}+4}{\cancel{2}} = 1 + 4 = 5$ is *NOT* correct because 3 cannot equal 5!

Therefore, NEVER CANCEL TERMS!

Note that 1 is always a factor of any expression, although it usually is not written. Thus, if all other factors are canceled, the factor 1 is still present. This is illustrated in the following example.

Example 4   Simplify: $\dfrac{x+4}{3x^2 + 5x - 28}$

Solution

$$\frac{\cancel{(x+4)}}{\cancel{(x+4)}\,(3x-7)}$$

$$\frac{1}{(3x-7)} \quad \blacksquare$$

Example 5   Simplify: $\dfrac{x^3 - 2x^2 - 15x}{x^3 - 3x^2 - 10x}$

Solution   Remember that factoring can require more than one step.

$$\frac{x\left(x^2 - 2x - 15\right)}{x\left(x^2 - 3x - 10\right)}$$

$$\frac{\cancel{x}\,\cancel{(x-5)}\,(x+3)}{\cancel{x}\,\cancel{(x-5)}\,(x+2)}$$

$$\frac{x+3}{x+2} \quad \blacksquare$$

Example 6   Simplify: $\dfrac{x^2 + x - 6}{x^2 + 3x + 2}$

Solution

$$\frac{(x+3)(x-2)}{(x+2)(x+1)}$$

There are no common factors; hence the expression is in simplest form. It is acceptable to write this expression in either factored form or expanded form. $\blacksquare$

Example 7   Simplify: $\dfrac{x-2}{2-x}$

Solution   Note that the factor $(2-x)$ is the negative of $(x-2)$. So $(2-x)$ can be written as $-(x-2)$ or $(-1)(x-2)$.

$$\frac{\cancel{(x-2)}}{-1\cancel{(x-2)}}$$

The simplified form is -1. ∎

The idea in Example 7 can be generalized. Any time a factor in the numerator is the opposite of a factor in the denominator, the result will be -1.

RULE

$$\frac{a-b}{b-a} = \frac{(a-b)}{-1(a-b)} = -1, \quad a \neq b$$

Example 8   Simplify: $\dfrac{6-x-x^2}{x^2-4}$

Solution

$$\frac{(3+x)(2-x)}{(x+2)(x-2)}$$

$$\frac{(3+x)\cancel{(x-2)}(-1)}{(x+2)\cancel{(x-2)}}$$

$$-\frac{3+x}{x+2} \quad \text{or} \quad -\frac{x+3}{x+2} \quad ∎$$

EXERCISE 5-1-1

Simplify when possible.

1. $\dfrac{x^2+2x}{x^2+3x+2}$

2. $\dfrac{x}{x^2+3x}$

3. $\dfrac{x^2-1}{x^2+x-2}$

4. $\dfrac{a+3}{a^2-9}$

5. $\dfrac{y^2+2y+1}{y^2-4y-5}$

6. $\dfrac{x^2-4}{x^2+4x+4}$

7. $\dfrac{x^2-4x+3}{x^2-6x+9}$

8. $\dfrac{p^2+4p+4}{p^2-4}$

9. $\dfrac{x^2+4x-12}{x^2-4x}$

10. $\dfrac{z^2-z-2}{z^2+z-2}$

11. $\dfrac{x^2+5x}{x^2+8x+15}$

12. $\dfrac{8-2y-y^2}{y+4}$

13. $\dfrac{a^2+12a+35}{a^2+4a-21}$

14. $\dfrac{16-9x^2}{3x^2-16x+16}$

15. $\dfrac{9-x^2}{x^2+x-12}$

16. $\dfrac{p^2+3p}{p^3+9p^2+18p}$

17. $\dfrac{x^3-2x^2-3x}{x^3+3x^2+2x}$

18. $\dfrac{3-5x-2x^2}{2x^2+9x-5}$

19. $\dfrac{3a^2+17ab+10b^2}{2a^2+7ab-15b^2}$

20. $\dfrac{2a^2+2ab+2b^2}{a^3-b^3}$

21. $\dfrac{27x^3+1}{3x^2-2x-1}$

5–2 MULTIPLICATION AND DIVISION OF RATIONAL EXPRESSIONS

OBJECTIVES

Upon completion of this section you should be able to:
1. Multiply two or more rational expressions.
2. Divide rational expressions.

Now that you have learned to simplify rational expressions, we wish to present the basic operations that can be performed on them. You will find that the operations of multiplication and division presented in this section are closely related to the process of simplifying.

Recall that the product of two fractions is equal to the product of their numerators divided by the product of their denominators.

$$\frac{a}{b} \cdot \frac{c}{d} = \frac{ac}{bd}$$

For instance,

$$\frac{2}{7} \cdot \frac{7}{12} = \frac{(2)(7)}{(7)(12)} = \frac{14}{84}.$$

The final answer would now need to be simplified. This means that 14 and 84 must be factored and the common factors cancelled.

$$\frac{14}{84} = \frac{(\not{7})(\not{2})}{(\not{7})(\not{2})(2)(3)} = \frac{1}{6}$$

From this example we can see that we may be able to shorten the multiplication process. It is a duplication of effort if we first multiply and then proceed to simplify. A more direct approach is to factor, cancel common factors, and then multiply the remaining factors.

$$\frac{\not{2}}{\not{7}} \cdot \frac{\not{7}}{\not{12}_{6}} = \frac{1}{6}$$

RULE

To multiply rational expressions:
1. Factor all numerators and denominators.
2. Find common factors and cancel.
3. Multiply the remaining factors in the numerator and place this product over the product of the remaining factors in the denominator.

Example 1   Multiply: $\dfrac{2x^2-5x-12}{x^2-3x-4}\cdot\dfrac{x^2-1}{2x^2+7x+6}$

Solution   Factor numerators and denominators.

$$\frac{(2x+3)(x-4)}{(x-4)(x+1)}\cdot\frac{(x+1)(x-1)}{(2x+3)(x+2)}$$

Find common factors and cancel.

$$\frac{\cancel{(2x+3)}\,\cancel{(x-4)}}{\cancel{(x-4)}\,\cancel{(x+1)}}\cdot\frac{\cancel{(x+1)}\,(x-1)}{\cancel{(2x+3)}\,(x+2)}$$

The answer is in simplest form.

$$\frac{x-1}{x+2}\quad\blacksquare$$

Example 2   Multiply: $\dfrac{x^2-6x+8}{x^2+5x-14}\cdot\dfrac{x^2-x-6}{8+2x-x^2}$

Solution

$$\frac{(x-4)(x-2)}{(x-2)(x+7)}\cdot\frac{(x+2)(x-3)}{(4-x)(2+x)}$$

$$\frac{\cancel{(x-4)}\,\cancel{(x-2)}}{\cancel{(x-2)}\,(x+7)}\cdot\frac{\cancel{(x+2)}\,(x-3)}{(-1)\,\cancel{(x-4)}\,\cancel{(2+x)}}$$

$$-\frac{(x-3)}{(x+7)}$$

This answer could have different forms.

$$-\frac{x-3}{x+7}\ \text{or}\ \frac{-(x-3)}{x+7}\ \text{or}\ \frac{-x+3}{x+7}\ \text{or}\ \frac{3-x}{x+7}\ \text{or}\ \frac{x-3}{-(x+7)}\ \text{or}\ \frac{x-3}{-x-7}$$

Any one of these forms is considered correct.   ∎

    The multiplication process can be extended to find the product of any number of rational expressions.

Example 3   Multiply: $\dfrac{x^2+4x+4}{x^2-x-6}\cdot\dfrac{2x-6}{4x-3}\cdot\dfrac{2x^2-7x+6}{8x^2+10x-12}$

Solution

$$\frac{(x+2)(x+2)}{(x+2)(x-3)}\cdot\frac{2(x-3)}{(4x-3)}\cdot\frac{(2x-3)(x-2)}{2(x+2)(4x-3)}$$

$$\frac{\cancel{(x+2)}\,\cancel{(x+2)}}{\cancel{(x+2)}\,\cancel{(x-3)}}\cdot\frac{\cancel{2}\,\cancel{(x-3)}}{(4x-3)}\cdot\frac{(2x-3)(x-2)}{\cancel{2}\,\cancel{(x+2)}\,(4x-3)}$$

$$\frac{(2x-3)(x-2)}{(4x-3)^2}\quad\blacksquare$$

Recall that division is defined as multiplication by the reciprocal of the divisor.

$$\frac{a}{b} \div \frac{c}{d} = \frac{a}{b} \cdot \frac{d}{c}$$

> **RULE**
>
> To divide rational expressions, multiply the first expression by the reciprocal of the second.

Example 4   Divide: $\dfrac{2x^2 + 7x + 5}{3x^2 - x - 14} \div \dfrac{2x^2 - x - 15}{x^2 - x - 6}$

Solution   Write the reciprocal of the divisor and multiply. $\dfrac{2x^2 + 7x + 5}{3x^2 - x - 14} \cdot \dfrac{x^2 - x - 6}{2x^2 - x - 15}$

Now proceed as in multiplication.

$$\frac{(2x+5)(x+1)}{(3x-7)(x+2)} \cdot \frac{(x+2)(x-3)}{(x-3)(2x+5)}$$

$$\frac{x+1}{3x-7} \qquad \blacksquare$$

Example 5   Divide: $\dfrac{6x^2 - 5x - 4}{3x^2 - 10x + 8} \div \left(2x^2 + 5x + 2\right)$

Solution   In this case the divisor, $2x^2 + 5x + 2$, can be written as $\dfrac{2x^2 + 5x + 2}{1}$.

$$\frac{6x^2 - 5x - 4}{3x^2 - 10x + 8} \cdot \frac{1}{2x^2 + 5x + 2}$$

$$\frac{(2x+1)(3x-4)}{(3x-4)(x-2)} \cdot \frac{1}{(2x+1)(x+2)}$$

$$\frac{1}{(x-2)(x+2)} \qquad \blacksquare$$

EXERCISE 5-2-1

Perform the indicated operation.

1. $\dfrac{2x+2}{x+3} \cdot \dfrac{x+3}{2x-6}$

2. $\dfrac{x^2 + x}{x-1} \cdot \dfrac{2x-2}{x+1}$

3. $\dfrac{a^2 - 3a}{a^2 - 9} \cdot \dfrac{a + 3}{a^2 + a}$

4. $\dfrac{y^2 + 2y + 1}{y^2 + y} \cdot \dfrac{1}{y + 1}$

5. $\dfrac{3x + 15}{x^2 + 6x + 5} \cdot \dfrac{x^2 + 2x + 1}{3x^2 - 3}$

6. $\dfrac{b^2 - 1}{5b - 5} \cdot \dfrac{5}{b^2 + 5b + 4}$

7. $\dfrac{m^2 + 4m + 3}{m - 5} \cdot \left(m^2 - 2m - 15 \right)$

8. $\dfrac{x^2 + 5x - 14}{x - 3} \cdot \dfrac{6 - 2x}{x^2 + 12x + 35}$

9. $\dfrac{x + 3}{6x^2 - 17x + 10} \cdot \dfrac{6x^2 - 11x + 5}{x^2 + 4x + 3}$

10. $\dfrac{1 - 2x}{3x^2 - 2x - 8} \cdot \dfrac{3x + 4}{2x^2 + x - 1}$

11. $\dfrac{12x^2 - 19x - 10}{x^2 + x - 6} \div \dfrac{x + 7}{x + 3}$

12. $\dfrac{x^2 - 4}{x^2 + 2x} \div \dfrac{x - 1}{x^2 - x}$

13. $\dfrac{x}{x - 1} \div \dfrac{x - 1}{2}$

14. $\dfrac{a - 1}{a} \div \dfrac{3}{a - 1}$

15. $\dfrac{t^2 + 8t + 15}{t^2 - 4t - 21} \div \dfrac{t + 5}{t^2 - 3t - 28}$

16. $\dfrac{a^2 - 2a - 3}{1 - a^2} \div \dfrac{a^2 - 5a + 6}{a^2 - a}$

17. $\dfrac{2p^2+3p+1}{p^2+2p+1} \div \dfrac{2p^2+p}{p}$

18. $\dfrac{3x^2+9x-54}{2x^2-2x-12} \div \dfrac{3x^2+21x+18}{4x^2-12x-40}$

19. $\dfrac{2x^2+9x+10}{x^2-7x-18} \div (2x+5)$

20. $\dfrac{25x^2-1}{10x^2+17x+3} \div (1-5x)$

21. $\dfrac{x^2+6x+5}{2x^2-2x-12} \cdot \dfrac{4x^2-36}{x^2+8x+15}$

22. $\dfrac{9x^2+9x-4}{6x^2-11x+3} \cdot (15-10x)$

23. $\dfrac{3x^2+11x+10}{x^2-4} \div \dfrac{2x^2+7x-4}{x^2+2x-8}$

24. $\dfrac{2x^2+11x+12}{4x^2+16x+15} \div \dfrac{2x^2+3x-20}{4x^2-25}$

25. $\dfrac{3x^2+10x+3}{x^2+10x+21} \cdot \dfrac{2x^2+9x-35}{6x^2-13x-5}$

26. $\dfrac{6+x-x^2}{2x^2+3x-2} \div \dfrac{x^2+4x-21}{3x^2+25x+28}$

27. $\dfrac{b^3+8}{b^2-4} \cdot \dfrac{b^2+4b+4}{b^2+2b}$

28. $\dfrac{m}{m^3-27} \div \dfrac{m^2-3m}{m^2+3m+9}$

29. $\dfrac{x+3}{x^2-4x-5} \cdot \dfrac{x^2-2x-3}{x^2+5x+6} \cdot \dfrac{x^2+x-30}{x^2+x-12}$

30. $\dfrac{2x^2+9x-5}{x^2+10x+21} \cdot \dfrac{3x^2+11x+6}{2x^2-13x+6} \cdot \dfrac{x^2+3x-28}{3x^2-10x-8}$

5–3 FINDING THE LEAST COMMON DENOMINATOR

> **OBJECTIVES**
>
> Upon completion of this section you should be able to:
> 1. Combine rational expressions having the same denominator.
> 2. Find the least common denominator (LCD) of several rational expressions.
> 3. Rewrite a rational expression as an equivalent expression having a different denominator.

Only like quantities can be added or subtracted. **Like fractions** are those that have the same denominator. Therefore, only like fractions can be added or subtracted. When two fractions have the same denominator, their sum or difference is the sum or difference of their numerators over their common denominator.

$$\frac{a}{b}+\frac{c}{b}=\frac{a+c}{b} \quad \text{and} \quad \frac{a}{b}-\frac{c}{b}=\frac{a-c}{b}$$

Example 1 Add: $\dfrac{1}{5}+\dfrac{3}{5}$

Solution Since these fractions have a common denominator, we put the sum of the numerators over the common denominator.

$$\frac{1+3}{5}=\frac{4}{5} \quad \blacksquare$$

Example 2 Add: $\dfrac{2}{c}+\dfrac{3}{c}+\dfrac{x}{c}$

Solution Since these rational expressions have a common denominator, we put the sum of the numerators over the common denominator.

$$\frac{2+3+x}{c}$$

$$\frac{5+x}{c} \quad \blacksquare$$

Example 3 Subtract: $\dfrac{2}{3}-\dfrac{x}{3}$

Solution Since these rational expressions have a common denominator, we put the difference of the numerators over the common denominator.

$$\frac{2-x}{3} \quad \blacksquare$$

Example 4   Subtract: $\dfrac{7}{8} - \dfrac{x+5}{8}$

Solution   Recall that the fraction bar is a grouping symbol. The *quantity* $(x + 5)$ must be subtracted.

$$\frac{7-(x+5)}{8}$$

$$\frac{7-x-5}{8}$$

$$\frac{2-x}{8} \quad \blacksquare$$

EXERCISE 5-3-1

Perform the indicated operations.

1. $\dfrac{1}{7} + \dfrac{4}{7}$

2. $\dfrac{2}{13} + \dfrac{8}{13}$

3. $\dfrac{3}{x} + \dfrac{5}{x}$

4. $\dfrac{7}{a} - \dfrac{2}{a}$

5. $\dfrac{1}{y} - \dfrac{a}{y}$

6. $\dfrac{b}{x} + \dfrac{1}{x}$

7. $\dfrac{3}{x} - \dfrac{4}{x} + \dfrac{a}{x}$

8. $\dfrac{2}{y} - \dfrac{c}{y} + \dfrac{4}{y}$

9. $\dfrac{a}{x} + \dfrac{b}{x} + \dfrac{c}{x}$

10. $\dfrac{a}{c} + \dfrac{3}{c} + \dfrac{a}{c}$

11. $\dfrac{6}{x} - \dfrac{x+2}{x}$

12. $\dfrac{5}{a} - \dfrac{a+4}{a}$

When adding or subtracting fractions that do not have the same denominators, it is necessary to rewrite them so that they have a common denominator. The fundamental principle of fractions gives us a way to do this. For instance, the fraction $\dfrac{2}{3}$ can be rewritten by multiplying the numerator and denominator by the same nonzero number.

$$\frac{2}{3} = \frac{2 \cdot 5}{3 \cdot 5} = \frac{10}{15}$$

$$\frac{2}{3} = \frac{2 \cdot 7}{3 \cdot 7} = \frac{14}{21}$$

The new denominator will always have the original denominator as a factor. The fraction $\frac{2}{3}$ could not be rewritten as a simple fraction having a denominator of 10, since 3 is not a factor of 10. A **simple fraction** is a fraction whose numerator and denominator are not fractional in form.

A common denominator for $\frac{2}{3}$ and $\frac{5}{8}$ would be a number that contains 3 and 8 as factors. Some possible common denominators would be 24, 48, 72, and so on, all of which are divisible by both 3 and 8. To keep the rational expression as simple as possible when adding, we want to use the *least common denominator*. For the fractions $\frac{2}{3}$ and $\frac{5}{8}$, 24 is the least common denominator.

DEFINITION

The **least common denominator (LCD)** of two or more rational expressions is the smallest expression that contains all of the denominators as factors.

Example 5   Find the LCD: $\dfrac{x+5}{x-2}, \dfrac{3x+4}{x+6}$

Solution   Since $(x - 2)$ will not factor and $(x + 6)$ will not factor, the least common denominator is the product of $(x - 2)$ and $(x + 6)$, or $(x - 2)(x + 6)$. ■

Example 6   Find the LCD: $\dfrac{2x-3}{6x^2-5x-4}, \dfrac{x+1}{3x^2-10x+8}$

Solution   Of course, the product

$$\left(6x^2 - 5x - 4\right)\left(3x^2 - 10x + 8\right)$$

is a common denominator, but it may not be the *least* common denominator. If the two denominators contain like factors, then the least common denominator will *not* be their product. Thus we must first factor the denominators.

$$\frac{2x-3}{(3x-4)(2x+1)} \quad \text{and} \quad \frac{x+1}{(3x-4)(x-2)}$$

We are looking for the least expression that contains both denominators as factors. We know that the first denominator must be contained in the common denominator. In other words, the common denominator must contain at least the factors $(3x - 4)$ and $(2x + 1)$.

Proposed LCD: $(3x - 4)(2x + 1)$

Now we look at the next denominator. Its factors must also be contained in the common denominator. This means we need to include any factors of the second denominator that are not already present in our proposed LCD. Since $(3x - 4)$ is already present, we need only to include $(x - 2)$.

Thus our least common denominator is $(3x - 4)(2x + 1)(x - 2)$. ■

Notice in each of the preceding examples that the factors of each denominator are contained in the common denominator. Also note that no other factors are present and therefore this makes it the least common denominator.

RULE

To find the least common denominator (LCD) of several fractions:
1. Factor each denominator.
2. Write the first denominator as the proposed common denominator.
3. By inspection, determine which factors of the second denominator are not already in the proposed denominator and include them.
4. Repeat step 3 for each fraction.

The resulting expression is the *least* common denominator.

Example 7   Find the LCD: $\dfrac{a+2}{a^2-3a-4}, \dfrac{a+1}{a^2-8a+16}$

Solution   Factor the denominators.

$$\frac{a+2}{(a-4)(a+1)}, \frac{a+1}{(a-4)(a-4)}$$

The LCD is $(a-4)(a+1)(a-4)$ or $(a-4)^2(a+1)$.

Note that the factor $(a-4)$ occurs twice in the second denominator. Therefore, $(a-4)$ must be included twice in the LCD. ■

EXERCISE 5-3-2

Find the LCD for the fractions in each of the following problems.

1. $\dfrac{1}{x}, \dfrac{1}{y}, \dfrac{3}{z}$

2. $\dfrac{1}{x+1}, \dfrac{1}{x+3}$

3. $\dfrac{1}{3a}, \dfrac{2}{9a^2 b}, \dfrac{1}{2b^2}$

4. $\dfrac{3}{2x^2}, \dfrac{7}{3xy^2}, \dfrac{x}{8y^2 z}$

5. $\dfrac{1}{x}, \dfrac{1}{x+2}$

6. $\dfrac{2}{y-3}, \dfrac{5}{y}$

7. $\dfrac{1}{a^2 - 2a}, \dfrac{5}{a^2}$

8. $\dfrac{3}{b^2 + 7b}, \dfrac{4}{b}$

9. $\dfrac{a}{a-3}, \dfrac{1}{a^2 - 9}$

10. $\dfrac{1}{x+3}, \dfrac{1}{x^2 - x - 12}$

11. $\dfrac{b^2}{(b-4)^3(b+3)^3}, \dfrac{b}{(b-4)^2(b+3)}$

12. $\dfrac{a}{(a+2)^2(a-5)^2}, \dfrac{6a}{(a+2)(a-5)^3}$

13. $\dfrac{3x+1}{2x^2 + 11x + 12}, \dfrac{x-6}{2x^2 + 5x - 12}$

14. $\dfrac{2x-1}{2x^2 - x - 15}, \dfrac{x+4}{4x^2 - 25}$

15. $\dfrac{y+1}{y^2 - y - 6}, \dfrac{y}{y^2 - 6y + 9}, \dfrac{2y-1}{y^2 - 2y - 8}$

Now that we have discussed how to find the least common denominator of rational expressions, we can proceed to the next step in the process of adding or subtracting. We will need to write each rational expression as an equivalent rational expression having the LCD as its denominator. Using the fundamental principle of fractions, we will multiply the numerator and denominator of the rational expression by carefully chosen factors so that the new denominator is the LCD.

Example 8   Rewrite $\dfrac{2}{3}$ as an equivalent fraction with a denominator of 12.

Solution   We would choose to multiply by $\dfrac{4}{4}$ since $3 \cdot 4 = 12$.

$$\dfrac{2 \cdot 4}{3 \cdot 4} = \dfrac{8}{12} \quad \blacksquare$$

Example 9   Rewrite $\dfrac{3x+1}{2x^2+11x+12}$ as an equivalent fraction with a denominator of

$(2x + 3)\,(x + 4)\,(x - 2)$.

Solution   Factor the original denominator.

$$\frac{3x+1}{(2x+3)(x+4)}$$

Compare the original denominator to the new denominator. The denominator of the original fraction needs to be multiplied by $(x - 2)$ to obtain the new denominator. Thus the original numerator must also be multiplied by $(x - 2)$.

$$\frac{3x+1}{(2x+3)(x+4)}$$

$$\frac{(3x+1)(x-2)}{(2x+3)(x+4)(x-2)}$$

$$\frac{3x^2-5x-2}{(2x+5)(x+4)(x-2)}$$

Generally, the denominator need not be expanded. ■

EXERCISE 5-3-3

Rewrite each fraction as an equivalent fraction having the given denominator.

1. $\dfrac{1}{a}$ Denominator: $a^2 b^2$

2. $\dfrac{7}{xy^2}$ Denominator: $x^2 y^2 z$

3. $\dfrac{1}{x+2}$ Denominator: $(x+2)(x-3)$

4. $\dfrac{x+1}{2x+1}$ Denominator: $(2x+1)(x-2)$

5. $\dfrac{1}{x}$ Denominator: $x(x+2)$

6. $\dfrac{2}{y+3}$ Denominator: $y(y+3)$

7. $\dfrac{2a-1}{3a+4}$ Denominator: $(a-6)(3a+4)$

8. $\dfrac{2x-5}{3x+4}$ Denominator: $(3x+4)^2$

9. $\dfrac{x-5}{x^2+2x-15}$ Denominator: $(x+5)(x-3)(x+2)$

10. $\dfrac{2x+1}{x^2-3x-28}$ Denominator: $(x-7)(x+3)(x+4)$

5–4   ADDING RATIONAL EXPRESSIONS

> **OBJECTIVE**
>
> Upon completion of this section you should be able to add rational expressions.

In the previous section you learned to find the least common denominator of two or more rational expressions and also to rewrite a rational expression as an equivalent expression with a different denominator. We will now use these skills to add rational expressions.

> **RULE**
>
> To add two or more rational expressions:
> 1. Find the least common denominator (LCD) of all expressions to be added.
> 2. Rewrite each expression as an equivalent expression having the LCD as its denominator.
> 3. Place the sum of the numerators over the LCD.
> 4. Clear all parentheses in the numerator.
> 5. Combine like terms in the numerator.
> 6. Simplify the answer, if possible.

Example 1   Add: $\dfrac{3x}{y}+\dfrac{x}{2z}+\dfrac{3}{8}$

Solution   The LCD for y, $2z$, and 8 is $8yz$.

Rewrite each expression to have the LCD as its denominator.

$$\frac{3x}{y}\cdot\frac{8z}{8z}+\frac{x}{2z}\cdot\frac{4y}{4y}+\frac{3}{8}\cdot\frac{yz}{yz}$$

$$\frac{24xz}{8yz}+\frac{4xy}{8yz}+\frac{3yz}{8yz}$$

Add the numerators and put the sum over the common denominator.

$$\frac{24xz+4xy+3yz}{8yz}$$

The answer is in simplest form. ∎

Example 2   Add: $\dfrac{x+3y}{12}+\dfrac{x-2y}{18}$

Solution   The LCD of 12 and 18 is 36. Rewrite the expressions and add.

$$\frac{3(x+3y)}{3(12)}+\frac{2(x-2y)}{2(18)}$$

$$\frac{3(x+3y)+2(x-2y)}{36}$$

$$\frac{3x+9y+2x-4y}{36}$$

$$\frac{5x+5y}{36} \quad \blacksquare$$

Example 3   Add: $\dfrac{2}{x-5}+\dfrac{x-23}{x^2-x-20}$

Solution   Factor the denominators. $\dfrac{2}{(x-5)}+\dfrac{x-23}{(x-5)(x+4)}$

The LCD is $(x-5)(x+4)$.

Rewrite and add. $\dfrac{2(x+4)}{(x-5)(x+4)}+\dfrac{x-23}{(x-5)(x+4)}$

$$\frac{2(x+4)+x-23}{(x-5)(x+4)}$$

$$\frac{2x+8+x-23}{(x-5)(x+4)}$$

$$\frac{3x-15}{(x-5)(x+4)}$$

$$\frac{3\cancel{(x-5)}}{\cancel{(x-5)}(x+4)}$$

$$\frac{3}{x+4} \quad \blacksquare$$

Example 4   Add: $\dfrac{x-4}{x^2-2x-15}+\dfrac{3}{x^2+5x+6}$

Solution   Factor the denominators. $\dfrac{x-4}{(x+3)(x-5)}+\dfrac{3}{(x+3)(x+2)}$

The least common denominator is $(x + 3)(x - 5)(x + 2)$.

$$\frac{(x-4)(x+2)}{(x+3)(x-5)(x+2)}+\frac{3(x-5)}{(x+3)(x+2)(x-5)}$$

$$\frac{(x-4)(x+2)+3(x-5)}{(x+3)(x-5)(x+2)}$$

$$\frac{x^2-2x-8+3x-15}{(x+3)(x-5)(x+2)}$$

$$\frac{x^2+x-23}{(x+3)(x-5)(x+2)} \quad \blacksquare$$

CAUTION   A common error is to cancel expressions that are not in factored form. Do not cancel when writing the numerators over the common denominator. *Expand* the numerators and collect like terms. Then factor and cancel common factors if possible.

EXERCISE 5-4-1

Add.

1. $\dfrac{1}{x}+\dfrac{1}{y}$

2. $\dfrac{1}{x}+\dfrac{1}{y}+\dfrac{1}{z}$

3. $\dfrac{1}{3}+\dfrac{2}{x+5}$

4. $\dfrac{2}{x}+\dfrac{1}{x+1}$

5. $\dfrac{1}{y}+\dfrac{y}{y+3}$

6. $\dfrac{2}{p+2}+\dfrac{3}{p}$

7. $\dfrac{3}{2c-6}+\dfrac{1}{6}$

8. $\dfrac{1}{2}+\dfrac{2}{4z+8}$

9. $\dfrac{1}{b+2}+\dfrac{2}{b^2-b-6}$

10. $\dfrac{2}{x+1}+\dfrac{4x}{x^2+4x+3}$

11. $\dfrac{2}{x+2}+\dfrac{1}{x^2+2x}$

12. $\dfrac{z+3}{z-6}+\dfrac{z-1}{z^2-6z}$

13. $\dfrac{3}{x^2-9}+\dfrac{2}{x^2+6x+9}$

14. $\dfrac{6}{5x+10}+\dfrac{2x}{x^2-3x-10}$

15. $\dfrac{1}{b+2}+\dfrac{1}{b+1}+\dfrac{1}{b-3}$

16. $\dfrac{3}{x+1}+\dfrac{2}{x-3}+\dfrac{5}{x+4}$

17. $\dfrac{2}{x-2}+\dfrac{3}{x+1}+\dfrac{x-8}{x^2-x-2}$

18. $\dfrac{5}{y+4}+\dfrac{2y-13}{y^2+y-12}+\dfrac{1}{y-3}$

19. $\dfrac{p+3}{2-p-p^2}+\dfrac{2p-1}{p+2}$

20. $\dfrac{t+2}{t+5}+\dfrac{t-3}{10-3t-t^2}$

21. $\dfrac{2p}{p^2+8p+16}+\dfrac{p-1}{16-p^2}$

Practice your skills.

Sometimes, more than one operation may appear in a problem. Recall the order of operations and the meaning of the parentheses in the following problems.

22. $\dfrac{3}{x}+\dfrac{5}{y}\cdot\dfrac{4y}{3x}$

23. $\dfrac{1}{x^2-x-2}\div\dfrac{1}{x+1}+\dfrac{3}{x-2}$

24. $\dfrac{1}{x^2-x-2}\div\left(\dfrac{1}{x+1}+\dfrac{3}{x-2}\right)$

5–5   SUBTRACTING RATIONAL EXPRESSIONS

> **OBJECTIVE**
>
> Upon completion of this section you should be able to subtract rational expressions.

Recall that subtraction of fractions, like addition, can only be performed if the fractions have the same denominators. When the denominators are the same, the difference is found by placing the difference of the numerators over the common denominator.

$$\frac{a}{c} - \frac{b}{c} = \frac{a-b}{c}$$

If the denominators are not the same, a common denominator must be found and each fraction changed to an equivalent fraction having the common denominator. As was true in addition, our work can be simplified by using the *least* common denominator.

> **RULE**
>
> To subtract one rational expression from another.
> 1. Find the least common denominator (LCD) of the expressions to be subtracted.
> 2. Rewrite each fraction as an equivalent fraction having the LCD as its denominator.
> 3. Place the difference of the numerators over the LCD.
> 4. Clear all parentheses in the numerator.
> 5. Combine like terms in the numerator.
> 6. Simplify the answer, if possible.

Example 1   Subtract: $\dfrac{x+3}{x+4} - \dfrac{5}{x+2}$

Solution   The least common denominator is $(x+4)(x+2)$.

$$\frac{(x+3)(x+2)}{(x+4)(x+2)} - \frac{5(x+4)}{(x+2)(x+4)}$$

$$\frac{(x+3)(x+2) - 5(x+4)}{(x+4)(x+2)}$$

Remember, you cannot cancel in this step. Expand and collect like terms.

$$\frac{x^2 + 5x + 6 - 5x - 20}{(x+4)(x+2)}$$

$$\frac{x^2 - 14}{(x+4)(x+2)} \quad ■$$

Example 2   Subtract: $\dfrac{2}{x-3} - \dfrac{x+7}{x^2-x-6}$

Solution   Factor the denominators. $\dfrac{2}{(x-3)} - \dfrac{x+7}{(x-3)(x+2)}$

The least common denominator is $(x-3)(x+2)$.

$$\frac{2(x+2)}{(x-3)(x+2)} - \frac{x+7}{(x-3)(x+2)}$$

$$\frac{2(x+2)-(x+7)}{(x-3)(x+2)}$$

$$\frac{2x+4-x-7}{(x-3)(x+2)}$$

$$\frac{\cancel{x-3}}{\cancel{(x-3)}(x+2)}$$

$$\frac{1}{x+2} \quad \blacksquare$$

CAUTION   Note that when subtracting, the numerator of the second fraction is placed in parentheses. This is necessary since the subtraction sign will change each sign of the factor $(x+7)$, giving $-x-7$.

Example 3   Subtract: $\dfrac{x+1}{x^2-x-6} - \dfrac{x-2}{2x-6}$

Solution   Factor the denominators. $\dfrac{x+1}{(x-3)(x+2)} - \dfrac{x-2}{2(x-3)}$

The least common denominator is $2(x-3)(x+2)$.

$$\frac{2(x+1)}{2(x-3)(x+2)} - \frac{(x-2)(x+2)}{2(x-3)(x+2)}$$

$$\frac{2x+2-(x^2-4)}{2(x-3)(x+2)}$$

$$\frac{2x+2-x^2+4}{2(x-3)(x+2)}$$

$$\frac{-x^2+2x+6}{2(x-3)(x+2)} \quad \blacksquare$$

CAUTION   Note that the subtraction sign in the numerator changed every sign in the product $(x-2)(x+2)$. A good practice is to perform the multiplication $(x-2)(x+2) = x^2 - 4$ and then change each sign. Don't attempt to multiply and change the signs at the same time or mistakes are likely to result.

EXERCISE 5-5-1

Subtract.

1. $\dfrac{3}{5} - \dfrac{4}{x}$

2. $\dfrac{2}{x} - \dfrac{4}{y}$

3. $\dfrac{x}{x+1} - \dfrac{2}{3}$

4. $\dfrac{5}{6} - \dfrac{2x}{3x-12}$

5. $\dfrac{4}{x} - \dfrac{6}{x+3}$

6. $\dfrac{1}{a+2} - \dfrac{1}{a+3}$

7. $\dfrac{2b}{b-3} - \dfrac{4}{b+5}$

8. $\dfrac{c}{c+1} - \dfrac{4}{c^2+c}$

9. $\dfrac{p}{p+2} - \dfrac{p+1}{p^2+2p}$

10. $\dfrac{2x-1}{x-1} - \dfrac{3x+2}{x^2+3x-4}$

11. $\dfrac{5x+1}{x+3} - \dfrac{4x-13}{x^2+x-6}$

12. $\dfrac{5t+3}{t^2-2t-15} - \dfrac{t+2}{t-5}$

13. $\dfrac{y+1}{3y^2-12} - \dfrac{y-3}{y+2}$

14. $\dfrac{z+4}{z^2-z-2} - \dfrac{z}{z^2+z-6}$

15. $\dfrac{2x}{x^2-1} - \dfrac{2x+4}{x^2+4x-5}$

Practice your skills.
Perform the indicated operations. Recall the order of operations and the meaning of parentheses.

16. $\dfrac{2}{x+5} + \dfrac{1}{x-4} - \dfrac{3}{x+3}$

17. $\dfrac{y}{y^2-4y-5} - \dfrac{2}{y+1} + \dfrac{2y-1}{y^2-7y+10}$

18. $\dfrac{1}{x^2-x-6} \div \dfrac{1}{x+2} - \dfrac{1}{x-3}$

19. $\dfrac{1}{a^2-a-6} \div \left(\dfrac{1}{a+2} - \dfrac{1}{a-3} \right)$

20. $\dfrac{2}{x-1} \cdot \dfrac{3x-3}{x+5} + \dfrac{x+1}{x^2+4x-5} \div \dfrac{2}{x-1} - \dfrac{1}{2}$

5–6   SIMPLIFYING COMPLEX FRACTIONS

OBJECTIVES

Upon completion of this section you should be able to:
1. Identify a complex fraction.
2. Simplify a complex fraction.

A fraction that has either a numerator or denominator, or both, containing fractions is called a **complex fraction**.   The following are examples of complex fractions.

$$\frac{\dfrac{a}{b}}{\dfrac{c}{d}}, \quad \frac{\dfrac{x}{y}+1}{x}, \quad \frac{\dfrac{1}{x}+\dfrac{1}{y}}{\dfrac{a}{b}+\dfrac{c}{d}}$$

A complex fraction is not considered to be in simplest form. There are two methods that can be used to simplify a complex fraction. The first method uses the fundamental principle of fractions.

RULE FOR METHOD 1

To simplify a complex fraction:
1. Identify the rational expressions in the numerator and denominator.
2. Find the least common denominator (LCD) of all these expressions.
3. Multiply the numerator and denominator of the complex fraction by the LCD.
4. If necessary, simplify the resulting expression.

We will begin with a numerical example.

Example 1   Simplify: $\dfrac{\dfrac{3}{4}+\dfrac{1}{3}}{5-\dfrac{3}{4}}$

Solution   Identify the denominators of the fractions in the numerator and denominator. $\dfrac{\dfrac{3}{4}+\dfrac{1}{3}}{5-\dfrac{3}{4}}$

The LCD of these denominators is 12. Multiply the numerator and denominator of the complex fraction by 12.

$$\frac{(12)\left(\frac{3}{4}+\frac{1}{3}\right)}{(12)\left(5-\frac{3}{4}\right)}$$

$$\frac{\overset{3}{\cancel{12}}\left(\frac{3}{\cancel{4}}\right)+\overset{4}{\cancel{12}}\left(\frac{1}{\cancel{3}}\right)}{12(5)-\overset{3}{\cancel{12}}\left(\frac{3}{\cancel{4}}\right)}$$

$$\frac{9+4}{60-9}$$

$$\frac{13}{51} \quad\blacksquare$$

Example 2   Simplify: $\dfrac{\dfrac{1}{x}+\dfrac{1}{y}}{\dfrac{1}{x}-\dfrac{1}{y}}$

Solution   Identify the denominators of the rational expressions in the numerator and denominator.

$$\frac{\dfrac{1}{x}+\dfrac{1}{y}}{\dfrac{1}{x}-\dfrac{1}{y}}$$

The least common denominator of all the fractions is xy. Multiply the numerator and denominator of the complex fraction by xy.

$$\frac{xy\left(\dfrac{1}{x}+\dfrac{1}{y}\right)}{xy\left(\dfrac{1}{x}-\dfrac{1}{y}\right)}$$

$$\frac{\cancel{x}y\left(\dfrac{1}{\cancel{x}}\right)+x\cancel{y}\left(\dfrac{1}{\cancel{y}}\right)}{\cancel{x}y\left(\dfrac{1}{\cancel{x}}\right)-x\cancel{y}\left(\dfrac{1}{\cancel{y}}\right)}$$

$$\frac{y+x}{y-x} \quad\blacksquare$$

Example 3   Simplify: $\dfrac{\dfrac{1}{a^2} - \dfrac{1}{b^2}}{a+b}$

Solution   The denominators of the rational expressions in the numerator are a^2 and b^2.

$$\frac{\dfrac{1}{a^2} - \dfrac{1}{b^2}}{a+b}$$

The least common denominator is $a^2 b^2$.

$$\frac{a^2 b^2 \left(\dfrac{1}{a^2} - \dfrac{1}{b^2} \right)}{a^2 b^2 (a+b)}$$

$$\frac{\cancel{a^2} b^2 \left(\dfrac{1}{\cancel{a^2}} \right) - a^2 \cancel{b^2} \left(\dfrac{1}{\cancel{b^2}} \right)}{a^2 b^2 (a+b)}$$

$$\frac{b^2 - a^2}{a^2 b^2 (a+b)}$$

$$\frac{(b-a)\cancel{(b+a)}}{a^2 b^2 \cancel{(a+b)}}$$

$$\frac{b-a}{a^2 b^2} \quad \blacksquare$$

Example 4   Simplify: $\dfrac{\dfrac{2}{x} - \dfrac{1}{x+2}}{x}$

Solution   The LCD is $x(x+2)$.

$$\frac{x(x+2)\left(\dfrac{2}{x} - \dfrac{1}{x+2} \right)}{x(x+2)(x)}$$

$$\frac{\cancel{x}(x+2)\left(\dfrac{2}{\cancel{x}} \right) - x\cancel{(x+2)}\left(\dfrac{1}{\cancel{x+2}} \right)}{x^2 (x+2)}$$

$$\frac{2(x+2) - x}{x^2 (x+2)}$$

$$\frac{2x+4-x}{x^2 (x+2)}$$

$$\frac{x+4}{x^2 (x+2)} \quad \blacksquare$$

The second method for simplifying complex fractions uses the order of operations. Since the fraction bar is a grouping symbol, we perform the operations in the numerator and denominator first, and then divide.

> **RULE FOR METHOD 2**
>
> To simplify a complex fraction:
> 1. Perform the operations necessary to change both numerator and denominator to simple fractions.
> 2. Divide the numerator by the denominator.

The following examples are the problems from Examples 1 through 4 now simplified using Method 2. After studying both methods, you can decide which works best for you.

Example 5   Simplify: $\dfrac{\dfrac{3}{4}+\dfrac{1}{3}}{5-\dfrac{3}{4}}$

Solution
$$\left(\frac{3}{4}+\frac{1}{3}\right)\div\left(5-\frac{3}{4}\right)$$

$$\left(\frac{9}{12}+\frac{4}{12}\right)\div\left(\frac{20}{4}-\frac{3}{4}\right)$$

$$\frac{13}{12}\div\frac{17}{4}$$

$$\frac{13}{\underset{3}{\cancel{12}}}\cdot\frac{\cancel{4}}{17}=\frac{13}{51} \quad\blacksquare$$

Example 6   Simplify: $\dfrac{\dfrac{1}{x}+\dfrac{1}{y}}{\dfrac{1}{x}-\dfrac{1}{y}}$

Solution
$$\left(\frac{1}{x}+\frac{1}{y}\right)\div\left(\frac{1}{x}-\frac{1}{y}\right)$$

$$\left(\frac{y}{y}\cdot\frac{1}{x}+\frac{x}{x}\cdot\frac{1}{y}\right)\div\left(\frac{y}{y}\cdot\frac{1}{x}-\frac{x}{x}\cdot\frac{1}{y}\right)$$

$$\frac{y+x}{xy}\div\frac{y-x}{xy}$$

$$\frac{y+x}{\cancel{xy}}\cdot\frac{\cancel{xy}}{y-x}$$

$$\frac{y+x}{y-x} \quad\blacksquare$$

Example 7   Simplify: $\dfrac{\dfrac{1}{a^2} - \dfrac{1}{b^2}}{a+b}$

Solution $\left(\dfrac{1}{a^2} - \dfrac{1}{b^2}\right) \div (a+b)$

$\left(\dfrac{b^2}{b^2} \cdot \dfrac{1}{a^2} - \dfrac{a^2}{a^2} \cdot \dfrac{1}{b^2}\right) \div (a+b)$

$\dfrac{b^2 - a^2}{a^2 b^2} \cdot \dfrac{1}{a+b}$

$\dfrac{(b + a)(b - a)}{a^2 b^2} \cdot \dfrac{1}{(a + b)}$

$\dfrac{b - a}{a^2 b^2}$ ■

Example 8   Simplify: $\dfrac{\dfrac{2}{x} - \dfrac{1}{x+2}}{x}$

Solution $\left(\dfrac{2}{x} - \dfrac{1}{x+2}\right) \div x$

$\left(\dfrac{2(x+2)}{x(x+2)} - \dfrac{1 \cdot x}{x(x+2)}\right) \div x$

$\dfrac{2x + 4 - x}{x(x+2)} \div x$

$\dfrac{x + 4}{x(x+2)} \cdot \dfrac{1}{x}$

$\dfrac{x + 4}{x^2(x+2)}$ ■

EXERCISE 5-6-1

Simplify using either method.

1. $\dfrac{\dfrac{1}{a}}{\dfrac{1}{a^2}}$

2. $\dfrac{\dfrac{1}{a}}{\dfrac{1}{b}}$

3. $\dfrac{\dfrac{1}{x}}{y}$

4. $\dfrac{\dfrac{1}{x}+\dfrac{1}{y}}{xy}$

5. $\dfrac{\dfrac{1}{r}-\dfrac{1}{s}}{rs}$

6. $\dfrac{\dfrac{1}{a}+\dfrac{1}{b}}{a+b}$

7. $\dfrac{\dfrac{2}{x}+\dfrac{2}{y}}{x+y}$

8. $\dfrac{\dfrac{1}{x}-\dfrac{1}{y}}{\dfrac{1}{y}+\dfrac{1}{z}}$

9. $\dfrac{\dfrac{2}{c}+\dfrac{3}{d}}{\dfrac{1}{c}-\dfrac{1}{d}}$

10. $\dfrac{\dfrac{3}{a}+\dfrac{1}{2}}{2a}$

11. $\dfrac{1+\dfrac{1}{a}}{ab}$

12. $\dfrac{\dfrac{1}{a}+\dfrac{1}{b}}{a^2-b^2}$

13. $\dfrac{2-\dfrac{1}{x}}{x+1}$

14. $\dfrac{3+\dfrac{1}{a}}{\dfrac{1}{a+3}}$

15. $\dfrac{\dfrac{a}{b}-\dfrac{b}{a}}{\dfrac{1}{a+b}}$

16. $\dfrac{\dfrac{1}{x+2}-\dfrac{1}{x}}{2}$

17. $\dfrac{\dfrac{1}{p+4}-\dfrac{1}{p}}{4}$

18. $\dfrac{\dfrac{2}{a-2}-\dfrac{2}{a}}{2}$

19. $\dfrac{\dfrac{1}{x^2-1}+1}{\dfrac{1}{x+1}-\dfrac{1}{x-1}}$

20. $\dfrac{\dfrac{1}{x-3}+\dfrac{1}{x^2-2x-3}}{\dfrac{1}{x+1}}$

5–7   RATIONAL EQUATIONS

OBJECTIVES

Upon completion of this section you should be able to:

1. Solve rational equations.
2. Determine when a rational equation does not have a solution.

In Chapter 2, we solved many equations containing fractions. However, all of those fractions contained denominators without variables. In this section, we will solve equations with rational expressions such as

$$\frac{1}{x} + \frac{1}{x-1} = \frac{2}{x}.$$

A PROCEDURE FOR SOLVING RATIONAL EQUATIONS

1. Find the least common denominator (LCD) of all fractions in the equation.
2. Multiply both sides of the equation by the LCD.
3. Combine similar terms on each side of the equation.
4. Add or subtract numbers on both sides of the equation to get the unknowns on one side and the numbers on the other.
5. Divide both sides of the equation by the coefficient of the unknown and simplify.
6. Substitute the result of step 5 in the original equation to check the solution.

Note that step 6 is *necessary* when solving equations that contain variables in the denominator. Sometimes the answer will not check even though an error has not been made in the procedure. This happens when the solution results in a denominator of the original equation being equal to zero. In such cases the original equation has no solution.

Example 1   Solve for x: $\dfrac{2}{x} + \dfrac{3}{2} = 2$

Solution   The least common denominator is $2x$.

$$(2x)\left(\frac{2}{x} + \frac{3}{2}\right) = (2x)(2)$$

$$(2x)\left(\frac{2}{x}\right) + (2x)\left(\frac{3}{2}\right) = (2x)(2)$$

$$4 + 3x = 4x$$

$$4 = x$$

Check $\dfrac{2}{4} + \dfrac{3}{2} = 2$

$\dfrac{1}{2} + \dfrac{3}{2} = 2$

$2 = 2$ ∎

In Example 1, the number 4 does not result in division by zero when substituted into the original equation, so if no mistakes were made in the procedure, 4 is the solution. However, it is a good practice to write out the complete check.

Example 2   Solve for x: $\dfrac{2x}{x+1} + \dfrac{1}{3x-2} = 2$

Solution   The least common denominator is $(3x-2)(x+1)$.

$$(3x-2)(x+1)\left(\frac{2x}{x+1} + \frac{1}{3x-2}\right) = (3x-2)(x+1)(2)$$

$$(3x-2)\,\cancel{(x+1)}\left(\frac{2x}{\cancel{x+1}}\right) + \cancel{(3x-2)}\,(x+1)\left(\frac{1}{\cancel{3x-2}}\right) = 2(3x-2)(x+1)$$

$$(3x-2)(2x) + (x+1)(1) = 2\left(3x^2 + x - 2\right)$$

$$6x^2 - 4x + x + 1 = 6x^2 + 2x - 4$$

$$-3x + 1 = 2x - 4$$

$$-5x = -5$$

$$x = 1$$

We now check to see if the number 1 will give a 0 when substituted into the denominators of the original equation. It does not, so the solution is $x = 1$. ∎

Example 3   Solve for x: $\dfrac{2}{x^2-1} - \dfrac{1}{x+1} = \dfrac{1}{x-1}$

Solution   The least common denominator is $(x+1)(x-1)$.

$$(x+1)(x-1)\left(\frac{2}{(x+1)(x-1)} - \frac{1}{x+1}\right) = (x+1)(x-1)\left(\frac{1}{x-1}\right)$$

$$\cancel{(x+1)}\,\cancel{(x-1)}\left(\frac{2}{\cancel{(x+1)}\,\cancel{(x-1)}}\right) - \cancel{(x+1)}\,(x-1)\left(\frac{1}{\cancel{x+1}}\right) = (x+1)\,\cancel{(x-1)}\left(\frac{1}{\cancel{x-1}}\right)$$

$$2 - (x-1) = x+1$$

$$2 - x + 1 = x+1$$

$$3 - x = x+1$$

$$-2x = -2$$

$$x = 1$$

When we substitute 1 into the denominators of the original equation, we see that two of the denominators will be 0. Thus $x = 1$ cannot be a solution of the original equation. We conclude that there is no solution to this equation. So we write "No solution". ∎

It is easy to confuse simplifying expressions and solving equations. We did not use equal signs in the simplification of expressions in this chapter; we used equal signs in equations. We did this to try to make the difference between simplifying an expression and solving an equation very clear. The next two examples are presented to compare and contrast simplifying an expression (Example 4) and solving an equation (Example 5).

Example 4   Add: $\dfrac{2}{x}+\dfrac{x}{x+1}$

Solution   This is a simplification. We are simplifying the expression by adding. We must find the LCD of the fractions, rewrite each fraction, and add.

$$\frac{2}{x}\cdot\frac{(x+1)}{(x+1)}+\frac{x}{x+1}\cdot\frac{x}{x}$$

$$\frac{2x+2+x^2}{x(x+1)}$$

The result is a *rational expression* that represents the sum of the two original rational expressions. ■

Example 5   Solve: $\dfrac{2}{x}+\dfrac{x}{x+1}=1$

Solution   This is an equation to solve for x. We want to eliminate the denominators. So we find the LCD and then multiply both sides of the equation by it.

$$x(x+1)\left(\frac{2}{x}+\frac{x}{x+1}\right)=x(x+1)(1)$$

$$x(x+1)\frac{2}{x}+x(x+1)\frac{x}{x+1}=x(x+1)$$

$$2x+2+x^2=x^2+x$$

$$x=-2$$

The result is a *value* that yields a true statement when substituted for x in the original equation. ■

CAUTION   Be careful to distinguish between adding fractions and solving equations. Confusion may arise because the least common denominator is used in both processes.

In Chapter 3 we solved word problems by translating them into algebraic equations. It is possible that the equation for a word problem will be a rational equation.

Example 6   A camera utilizes the formula (from optics) $\dfrac{1}{f} = \dfrac{1}{a} + \dfrac{1}{b}$, where f represents the focal length of the lens, a represents the distance of an object from the lens, and b represents the distance of the image from the lens. If a lens has a focal length of 12 centimeters, how far from the lens will the image appear when the object is 36 centimeters from the lens?

Solution   Since the focal length is 12 cm and the object is 36 cm from the lens, let $f = 12$ and $a = 36$.

$$\frac{1}{12} = \frac{1}{36} + \frac{1}{b}$$

Solve for b. Multiply both sides of the equation by the LCD, $36b$.

$$36b\left(\frac{1}{12}\right) = 36b\left(\frac{1}{36} + \frac{1}{b}\right)$$
$$3b = b + 36$$
$$2b = 36$$
$$b = 18$$

The image will appear to be 18 cm from the lens. ∎

A category of problems known as **work problems** usually produces rational equations. The next two examples are work problems.

Example 7   Sara and Megan are roommates. They usually take turns cleaning their dorm room. Sara can clean the room in 2 hours. It takes Megan 3 hours to do the cleaning. If they worked together, how long would it take them to clean their room?

Solution   Let t represent the time it takes to clean the room when Sara and Megan work together. If Sara can clean the room in 2 hours, then she can clean $\dfrac{1}{2}$ of the room in one hour. If Megan can clean the room in 3 hours, she can clean $\dfrac{1}{3}$ of the room in one hour.

	Hours to complete the job	Amount of job completed in one hour
Sara	2	$\dfrac{1}{2}$
Megan	3	$\dfrac{1}{3}$
Together	t	$\dfrac{1}{t}$

The sum of the amount of work Sara and Megan can each do in one hour is equal to the total amount of work done in the hour.

$$\overbrace{\frac{1}{2}}^{\substack{\text{amount cleaned by} \\ \text{Sara in one hour}}} + \overbrace{\frac{1}{3}}^{\substack{\text{amount cleaned by} \\ \text{Megan in one hour}}} = \overbrace{\frac{1}{t}}^{\substack{\text{amount both clean} \\ \text{in one hour}}}$$

$$6t\left(\frac{1}{2}+\frac{1}{3}\right) = 6t\left(\frac{1}{t}\right)$$

$$3t + 2t = 6$$

$$5t = 6$$

$$t = \frac{6}{5}$$

This value for t does not give a zero denominator, so together Sara and Megan can clean their room in $1\frac{1}{5}$ hours (1 hour and 12 minutes). ∎

Example 8   Bill can clean the outside of a travel trailer in 6 hours. Bill and his helper working together can do the job in 4 hours. How long would it take his helper working alone to clean the trailer?

Solution   Let x represent the amount of time the helper takes to clean the trailer alone.

	Hours to complete the job	Amount of job completed in one hour
Bill	6	$\frac{1}{6}$
Helper	x	$\frac{1}{x}$
Together	4	$\frac{1}{4}$

The sum of the amount of work Bill and his helper can each do in one hour is equal to the total amount of work done in the hour.

$$\overbrace{\frac{1}{6}}^{\substack{\text{amount cleaned by} \\ \text{Bill in one hour}}} + \overbrace{\frac{1}{x}}^{\substack{\text{amount cleaned by} \\ \text{helper in one hour}}} = \overbrace{\frac{1}{4}}^{\substack{\text{amount both clean} \\ \text{in one hour}}}$$

$$12x\left(\frac{1}{6}+\frac{1}{x}\right) = 12x\left(\frac{1}{4}\right)$$

$$2x + 12 = 3x$$

$$12 = x$$

This value for x does not give a zero denominator, so Bill's helper would take 12 hours to clean the trailer alone. ∎

EXERCISE 5-7-1

Solve.

1. $\dfrac{5}{x} = \dfrac{1}{2}$

2. $\dfrac{1}{x} + \dfrac{1}{2x} = 3$

3. $\dfrac{2}{a} + \dfrac{1}{3} = \dfrac{1}{2a} - 1$

4. $\dfrac{x+1}{x} - \dfrac{x-2}{2x} = 1$

5. $\dfrac{3}{c-1} = 1$

6. $\dfrac{t}{t-2} = \dfrac{4}{3}$

7. $\dfrac{2}{b} = \dfrac{1}{b+1}$

8. $\dfrac{x}{x+1} = \dfrac{x-1}{x-2}$

9. $\dfrac{1}{x+1} = \dfrac{2}{1-x^2}$

10. $\dfrac{t+1}{t^2+2t-3} + \dfrac{1}{t-1} = \dfrac{1}{t+3}$

11. $\dfrac{x+2}{x^2+7x} = \dfrac{1}{x+3}$

12. $\dfrac{1}{x^2-3x} = \dfrac{2}{x^2-9}$

13. $\dfrac{p}{p+1} + \dfrac{1}{2p+1} = 1$

14. $\dfrac{3}{z+5} = 1 - \dfrac{z-4}{2z+10}$

15. $\dfrac{2}{b-1} + \dfrac{b}{b+1} = 1 - \dfrac{1}{b^2-1}$

16. $\dfrac{2}{3a+6} = \dfrac{1}{6} - \dfrac{1}{2a+4}$

17. $\dfrac{x+1}{x+5} - \dfrac{2x+1}{x-2} = \dfrac{5-x^2}{x^2+3x-10}$

18. A camera utilizes the formula (from optics) $\dfrac{1}{f} = \dfrac{1}{a} + \dfrac{1}{b}$, where f represents the focal length of the lens, a represents the distance of the object from the lens, and b represents the distance of the image from the lens. If a lens has a focal length of 10 centimeters, how far from the lens will the image appear when the object is 50 centimeters from the lens?

19. Joe and Theresa want to paint their living room. From past experience they know that Joe can paint the room in 6 hours, and Theresa can paint it in 10 hours. How long will it take them if they work together?

20. If an electric circuit is wired in parallel with two resistors, the total resistance R (measured in ohms) is given by the formula $\dfrac{1}{R} = \dfrac{1}{r_1} + \dfrac{1}{r_2}$, where r_1 and r_2 represent the resistances of the two resistors. If $R = 300$ ohms and $r_2 = 750$ ohms, find r_1.

21. Mower A can cut a field in 6 hours and mower B can cut the same field in 2 hours. How long would it take both mowers working together to cut the field?

22. An experienced bricklayer and his apprentice can build a wall together in 3 hours. It would take the apprentice 12 hours to do the job alone. How long would it take the experienced bricklayer to do the job alone?

23. Donna, Linda, and Dinah have pieced a quilt top. They plan to work together to stitch the quilt. Donna can usually finish stitching a quilt in 20 hours, while Linda and Dinah can each finish in 30 hours. How long will it take to finish the quilt if they all work together?

24. Two pumps working together can fill a swimming pool in 4 hours. If pump A can fill it alone in 5 hours, how long would it take pump B to fill it alone?

25. Harding, Marcel, and Nagesh are designing a new algebra book. Working alone, Marcel can prepare a chapter in 12 days. Nagesh can prepare the same chapter in 8 days. When all three people work together, they can finish the job in 3 days. How many days would Harding need to prepare the chapter working alone?

CHAPTER 5 SUMMARY

The number in brackets refers to the section of the chapter that discusses the concept.

Terminology

- A **rational expression** is the quotient of two polynomials. [5–1]
- The **fundamental principle of fractions** is $\dfrac{a}{b} = \dfrac{ax}{bx}$, $x \neq 0, b \neq 0$. [5–1]
- A fraction is in **simplest form** if the numerator and denominator have no factors in common. [5–1]
- The **least common denominator** (LCD) of two or more expressions is the smallest expression that contains all of the denominators as factors. [5–3]
- A **complex fraction** is a fraction that has either a numerator or denominator, or both, composed of fractions. [5–6]

Rules and Procedures

Simplifying Rational Expressions

- To simplify a rational expression, factor both the numerator and denominator, then cancel all common factors. [5–1]

Multiplication of Rational Expressions

- To multiply rational expressions: [5–2]
  1. Factor all numerators and denominators.
  2. Cancel any factor that is common to a numerator and a denominator
  3. Multiply the remaining factors in the numerator and place this product over the product of the remaining factors in the denominator.

Division of Rational Expressions

- To divide rational expressions, multiply by the reciprocal of the divisor. [5–2]

Least Common Denominator

- To find the least common denominator (LCD) of two or more rational expressions: [5–3]
  1. Factor each denominator.
  2. Write the first denominator as the proposed common denominator.
  3. By inspection, determine which factors of the second denominator are not already in the proposed denominator and include them.
  4. Repeat step 3 for each fraction.
  The resulting expression is the *least* common denominator.

Addition and Subtraction of Rational Expressions

- To add or subtract rational expressions, rewrite all expressions so they have a common denominator and combine the numerators over the LCD. [5–4, 5–5]

Simplifying Complex Fractions

- Method 1. To simplify a complex fraction: [5–6]
  1. Identify the rational expressions in the numerator and denominator.
  2. Find the LCD (least common denominator) of all these expressions.
  3. Multiply the numerator and denominator of the complex fraction by the LCD.
  4. If necessary, simplify the resulting expression.

- Method 2.   To simplify a complex fraction: [5–6]
    1. Perform the operations necessary to change both numerator and denominator to simple fractions.
    2. Divide the numerator by the denominator.

Rational Equations

- To solve equations containing rational expressions: [5–7]
    1. Find the LCD (least common denominator) of all fractions in the equation.

2. Multiply both sides of the equation by the LCD.
3. Combine similar terms on each side of the equation.
4. Add or subtract numbers on both sides of the equation to get the unknowns on one side and the numbers on the other.
5. Divide both sides of the equation by the coefficient of the unknown and simplify.
6. Substitute the result of step 5 in the original equation to check the solution.

CHAPTER 5 REVIEW

Simplify.

1. $\dfrac{3x+45}{12}$

2. $\dfrac{2x+8}{3x+12}$

3. $\dfrac{y^2-2y-15}{y^2+3y-40}$

4. $\dfrac{a^2-49}{a^2+14a+49}$

5. $\dfrac{6x+30}{10x^2+40x-50}$

Find the least common denominator.

6. $\dfrac{3}{x}, \dfrac{2}{x+2}$

7. $\dfrac{4}{x^2+3x}, \dfrac{x}{x+3}$

8. $\dfrac{3}{y^2-5y}, \dfrac{y-1}{y^2-4y-5}$

9. $\dfrac{4}{a+1}, \dfrac{1}{a^2-4}, \dfrac{5}{a+2}$

10. $\dfrac{20}{x^2-1}, \dfrac{3}{x^2-x-2}$

Perform the indicated operation.

11. $\dfrac{3x+6}{x} \cdot \dfrac{5x}{21x+42}$

12. $\dfrac{t^2-25}{t^2+7t+10} \cdot \dfrac{t^2+t-2}{t^2-9t+20}$

13. $\dfrac{3x^2+4x+1}{1-x^2} \cdot \dfrac{x^2-2x+1}{3x^2+x}$

14. $\dfrac{x^2+3x+2}{x^2+5x+6} \cdot \dfrac{x^2-2x-15}{x^2-3x-10} \cdot \dfrac{2x^2-x-10}{x^2+5x+4}$

15. $\dfrac{2x-6}{3x} \div \dfrac{x^2-2x-3}{x^2+x}$

16. $\dfrac{x+1}{x^2-5x-6} \div \dfrac{x^2-1}{6-x}$

17. $\dfrac{x^2+7x+10}{x^2+4x-5} \div \dfrac{x^2+7x+12}{x^2+2x-3}$

18. $\dfrac{9x^2+18x+8}{6x^2+19x+10} \div \dfrac{12x^2+13x-4}{8x^2+10x-3}$

19. $\dfrac{3}{x}+\dfrac{2}{x+2}$

20. $\dfrac{4}{t^2+3t}+\dfrac{t}{t+3}$

21. $\dfrac{3}{x^2-5x}+\dfrac{x-1}{x^2-4x-5}$

22. $\dfrac{4}{b+1}+\dfrac{1}{b^2-4}+\dfrac{5}{b+2}$

23. $\dfrac{2x}{x^2-1}+\dfrac{3}{x^2-x-2}$

24. $\dfrac{3}{4}-\dfrac{x}{x+5}$

25. $\dfrac{3}{s+1}-\dfrac{2}{s-2}$

26. $\dfrac{5}{b-1}-\dfrac{2b+6}{b^2+2b-3}$

27. $\dfrac{2x}{x^2-4}-\dfrac{x+4}{x^2+4x-12}$

28. $\dfrac{1}{p+2}+\dfrac{1}{p+1}-\dfrac{2}{p-7}$

Simplify.

29. $\dfrac{\dfrac{1}{a}+\dfrac{1}{b}}{a+b}$

30. $\dfrac{1-\dfrac{1}{a}}{a-1}$

31. $\dfrac{\dfrac{1}{x}-2}{\dfrac{1}{x+1}}$

32. $\dfrac{\dfrac{1}{x+2}+\dfrac{1}{x}}{2}$

33. $\dfrac{\dfrac{1}{x}+\dfrac{1}{y}}{\dfrac{1}{x+y}}$

Solve.

34. $\dfrac{3}{2x}=\dfrac{1}{x}+\dfrac{1}{2}$

35. $\dfrac{a}{a+7}=\dfrac{a-1}{a}$

36. $\dfrac{1}{x-2}+\dfrac{1}{x+2}=\dfrac{4}{x^2-4}$

37. $\dfrac{1}{x+1}-\dfrac{1}{2}=\dfrac{1}{3x+3}$

38. $\dfrac{x+1}{x+2}-\dfrac{x-3}{x+5}=\dfrac{4}{x^2+7x+10}$

39. In the electrical formula $\dfrac{1}{R} = \dfrac{1}{r_1} + \dfrac{1}{r_2} + \dfrac{1}{r_3}$, $R, r_1, r_2,$ and r_3 represent resistances (which are measured in ohms). Find r_2 if $R = 5$ ohms, $r_1 = 40$ ohms, and $r_3 = 8$ ohms.

40. A painter can paint a house alone in 12 hours. His helper can paint the same house alone in 15 hours. How long will it take both of them working together to paint the house?

CHAPTER 5 PRACTICE TEST

1. Simplify: $\dfrac{5x+15}{20}$

1. _____

2. Simplify: $\dfrac{a^2-2a-15}{a^2-25}$

2. _____

3. Multiply: $\dfrac{x^2+8x+15}{x^2+x-6}\cdot\dfrac{x^2+3x-10}{x^2+4x-5}$

3. _____

4. Divide: $\dfrac{2t+8}{3t}\div\dfrac{t^2+3t-4}{t^2-t}$

4. _____

5. Find the LCD: $\dfrac{4}{x},\ \dfrac{2}{x-7}$

5. _____

6. Find the LCD: $\dfrac{2}{x+6},\ \dfrac{5}{x^2-36},\ \dfrac{1}{x+4}$

6. _____

7. Add: $\dfrac{3}{x+1}+\dfrac{2}{x}$

7. _____

8. Subtract: $\dfrac{3z}{z+2} - \dfrac{z+3}{z^2+6z+8}$

8. _____

9. Simplify: $\dfrac{1+\dfrac{3}{a}}{a+3}$

9. _____

10. Simplify: $\dfrac{\dfrac{3}{y+2} - \dfrac{3}{y}}{2}$

10. _____

11. Solve: $\dfrac{3}{x+3} + \dfrac{1}{x-1} = \dfrac{8}{x^2+2x-3}$

11. _____

12. Water tap A can fill a washtub in 2 minutes. Water tap B can fill the same tub in 4 minutes. If both taps are open, how long will it take to fill the tub?

12. _____

CHAPTER 6

SURVEY

The following questions refer to material discussed in this chapter. Work as many problems as you can and check your answers with the answer key in the back of the book. The results will direct you to the sections of the chapter in which you need to work. If you answer all questions correctly, you may already have a good understanding of the material contained in this chapter.

1. Simplify: $\left(xy^2\right)^4$

1. _____

2. Simplify: $\left(\dfrac{x^2}{2}\right)^3\left(\dfrac{4}{x^3}\right)^2$

2. _____

Simplify; write final answers with positive exponents.

3. $\dfrac{x^{-2}y^{-5}}{x^{-4}y}$

3. _____

4. $\left(-5x^2\right)^{-3}\left(2x^{-2}\right)^2$

4. _____

5. Write the number 3,500,000 in scientific notation.

5. _____

6. Evaluate: $\sqrt[3]{-64}$

6. _____

246

7. Evaluate: $(-27)^{\frac{2}{3}}$

7. _____

8. Simplify: $\sqrt[3]{x^9 y^6}$

8. _____

9. Simplify: $x^{\frac{1}{2}} \cdot x^{\frac{2}{3}}$

9. _____

10. Simplify: $-2\sqrt{75}$

10. _____

Multiply.

11. $\sqrt{5}\left(3\sqrt{2} + \sqrt{3}\right)$

11. _____

12. $\left(2\sqrt{3} + 5\sqrt{7}\right)\left(2\sqrt{3} - 5\sqrt{7}\right)$

12. _____

13. Simplify: $\dfrac{3}{\sqrt[3]{x}}$

13. _____

14. Simplify: $\dfrac{4}{\sqrt{5} + \sqrt{2}}$

14. _____

CHAPTER 6

Exponents and Radicals

In this chapter we will continue to study the laws of exponents. We will also introduce you to techniques for simplifying radicals and the relationship that exists between radicals and exponents.

6–1   THE LAWS OF EXPONENTS

OBJECTIVES

Upon completion of this section you should be able to:
1. State the laws of exponents.
2. Apply one or more of the laws to simplify an expression.

In Chapter 1 we defined a positive integer exponent as a symbol that tells you how many times to use the base as a factor. We also gave the first two laws of exponents.

FIRST LAW OF EXPONENTS

$$x^a \cdot x^b = x^{a+b}$$

SECOND LAW OF EXPONENTS

If $x \neq 0$,

$$\frac{x^a}{x^b} = x^{a-b} \text{ if } a > b;$$

$$\frac{x^a}{x^b} = \frac{1}{x^{b-a}} \text{ if } a < b;$$

$$\frac{x^a}{x^b} = 1 \text{ if } a = b.$$

Remember that, to use these laws, the bases must be the same.

EXERCISE 6-1-1

As a review of the first two laws of exponents, simplify the following expressions.

1. $x^3 \cdot x^5$ 2. $x^2 \cdot x^4$ 3. $x \cdot x^6 \cdot x^3$ 4. $x^2 \cdot y^3$

5. $x^3 \cdot y \cdot x \cdot y^2$ 6. $\dfrac{x^5}{x^3}$ 7. $\dfrac{x^4}{x}$ 8. $\dfrac{x^2}{y}$

9. $\dfrac{x^3}{x^4}$ 10. $\dfrac{x^2 y^3}{x^3 y}$

There are three more laws of exponents. For each, we will give the law, illustrate why it works, and give examples demonstrating how to use it in simplifying expressions.

THIRD LAW OF EXPONENTS

$$\left(x^a\right)^b = x^{ab}$$

(To find a power of a power, keep the base and multiply the exponents.)

To illustrate this law, consider the expression $\left(x^5\right)^2$. The exponent 2 tells us to use the base x^5 as a factor two times. So we have $\left(x^5\right)^2 = x^5 \cdot x^5$. Now, since we are multiplying expressions with like bases, we can add exponents, $x^{5+5} = x^{10}$. This gives the same result as multiplying the original exponents, $\left(x^5\right)^2 = x^{5 \cdot 2} = x^{10}$.

Example 1   Simplify: $\left(x^4\right)^3$

Solution   We are cubing a variable already raised to the fourth power, so we keep the base and multiply the exponents.

$x^{4 \cdot 3}$

x^{12} ■

Example 2   Simplify: $\left(2^3\right)^2$

Solution $2^{3 \cdot 2}$

2^6 ■

Example 3   Simplify: $\left[\left(y^2\right)^3\right]^2$

Solution $\left[y^6\right]^2$

y^{12} ■

FOURTH LAW OF EXPONENTS

$$(xy)^a = x^a y^a$$

(Raising the product of factors to a power is the same as raising each factor to the power and then finding the product.)

Consider $(xy)^3$. The exponent, 3, tells us to use the base xy as a factor three times, $(xy)(xy)(xy)$. This simplifies to $x^3 y^3$. So the fourth law allows us to write $(xy)^3$ as $x^3 y^3$.

Example 4   Rewrite: $(ab)^5$

Solution   The expression consists of factors raised to a power. So the above rule applies.

$a^5 b^5$ ■

Example 5   Rewrite: $\left[(2)(3)\right]^2$

Solution $(2)^2 (3)^2$ ■

Simplification of expressions with exponents may require a combination of the laws of exponents. Keep this in mind as you read the following examples.

Example 6   Simplify: $\left(x^2 y^3\right)^2$

Solution   Using the fourth law, we raise the factors x^2 and y^3 to the second power.

$$\left(x^2\right)^2 \left(y^3\right)^2$$

Now using the third law, we have

$x^4 y^6$. ■

If you are unsure where to begin the simplification, recall the order of operations.

Example 7   Simplify: $\left[-2\left(a^2 b^5 \right)^2 \right]^3$

Solution   We will eliminate the innermost parentheses first. Note that there is an exponent and a multiplication. Since exponentiation is to be done first, we will use the fourth law.

$$\left[-2\left(a^2 \right)^2 \left(b^5 \right)^2 \right]^3$$

Next use the third law.

$$\left[-2a^4 b^{10} \right]^3$$

Apply the fourth and third laws again, respectively.

$$(-2)^3 \left(a^4 \right)^3 \left(b^{10} \right)^3$$
$$-8a^{12} b^{30} \quad \blacksquare$$

Example 8   Simplify: $\left(-x^4 y^3 \right)^2$

Solution $$(-1)^2 \left(x^4 \right)^2 \left(y^3 \right)^2$$
$$x^8 y^6 \quad \blacksquare$$

CAUTION   Do not attempt to use the fourth law of exponents on sums or differences since

$$(x+y)^2 = (x+y)(x+y) = x^2 + 2xy + y^2,$$
$$\text{NOT } x^2 + y^2.$$

FIFTH LAW OF EXPONENTS

$$\left(\frac{x}{y} \right)^a = \frac{x^a}{y^a}, y \neq 0$$

(Raising a quotient to a power is the same as raising the numerator and denominator to the power and then finding the quotient.)

To illustrate this fifth law, consider the expression $\left(\dfrac{x}{y} \right)^4$. The exponent, 4, means we are to use the base $\dfrac{x}{y}$ as a factor four times, $\left(\dfrac{x}{y} \right)\left(\dfrac{x}{y} \right)\left(\dfrac{x}{y} \right)\left(\dfrac{x}{y} \right)$. We know that to multiply fractions, we multiply the numerators and multiply the denominators, so this becomes $\dfrac{x^4}{y^4}$.

Example 9   Simplify: $\left(\dfrac{a}{2}\right)^2$

Solution   Here we are raising a quotient (fraction) to a power, so we use the fifth law.

$$\dfrac{a^2}{2^2} \text{ or } \dfrac{a^2}{4} \quad \blacksquare$$

Example 10   Simplify: $\left(\dfrac{x^2}{y^3}\right)^4$

Solution   To simplify we will need to use a combination of laws. We will use the fifth law and then the third law.

$$\dfrac{\left(x^2\right)^4}{\left(y^3\right)^4}$$

$$\dfrac{x^8}{y^{12}} \quad \blacksquare$$

Example 11   Simplify: $\left(\dfrac{-3x}{2y^2}\right)^3$

Solution   To simplify this expression we need to use the fifth, fourth, and third laws.

$$\dfrac{\left(-3\right)^3 x^3}{2^3 y^6}$$

$$\dfrac{-27x^3}{8y^6} \quad \blacksquare$$

Example 12   Simplify: $\left(3xy\right)^2\left(2x^2y^3\right)^3$

Solution $$\left(9x^2y^2\right)\left(8x^6y^9\right)$$

$$72x^8y^{11} \quad \blacksquare$$

Example 13   Simplify: $\left(\dfrac{x^4}{4}\right)^2\left(\dfrac{2}{x^2y}\right)^3$

Solution $$\left(\dfrac{x^8}{16}\right)\left(\dfrac{8}{x^6y^3}\right)$$

$$\dfrac{x^2}{2y^3} \quad \blacksquare$$

EXERCISE 6-1-2

Apply either one or a combination of the five laws of exponents.

1. $\left(x^2\right)^5$

2. $\left(ab\right)^3$

3. $\left(x^2y\right)^4$

4. $\left(\dfrac{x}{y}\right)^2$

5. $\left(\dfrac{m^2}{n^3}\right)^3$

6. $\left(\dfrac{x}{2}\right)^3$

7. $\left(-2a^2\right)^4$

8. $\left(-3xy^2\right)^3$

9. $\left(\dfrac{2}{x^2y}\right)^3$

10. $\left[\left(p^2\right)^3\right]^4$

11. $\left(\dfrac{xy^2}{x^2y}\right)^2$

12. $\left(\dfrac{2c^3}{cd^2}\right)^3$

13. $\left[2\left(xy\right)^2\right]^3$

14. $\left(\dfrac{-1}{5x^2}\right)^2$

15. $\left[\left(-2a^3b\right)^2\right]^3$

16. $\dfrac{\left(b^2\right)^3}{\left(b^3\right)^2}$

17. $\dfrac{\left(-3x\right)^3}{9x}$

18. $\dfrac{-4x^2}{\left(2x\right)^3}$

19. $\dfrac{-5xy^3}{\left(-5x\right)^2}$

20. $\dfrac{\left(2x^2y^3\right)^3}{\left(2x^3y^2\right)^2}$

21. $-2s^2t\left(s^2t\right)^3$

22. $\left(2xy\right)^3\left(2x^2y\right)^2$

23. $\left(-x^2y\right)^3\left(-2xy^3\right)^2$

24. $\left(x^2yz^2\right)^3\left(-2xy^3z\right)^2$

25. $\left[2\left(xy\right)^2\right]^5\left[3\left(x^2y\right)^3\right]^2$

26. $\left(6mn^2\right)\left(-4m^2n\right)\left(5m^3n^2\right)$

27. $\left(\dfrac{-2}{x}\right)^3\left(\dfrac{x}{2}\right)^2$

28. $\left(\dfrac{x^2y}{3}\right)\left(\dfrac{3}{xy^2}\right)^2$

29. $\left(\dfrac{-2x}{5y^2}\right)^3\left(\dfrac{5y}{4x^2}\right)^2\left(2x^5\right)$

30. $\left(\dfrac{x^3}{8}\right)^2\left(\dfrac{4}{x^2}\right)^3$

6–2   INTEGER EXPONENTS AND SCIENTIFIC NOTATION

OBJECTIVES

Upon completion of this section you should be able to:
1. Simplify expressions containing integer exponents.
2. Write a given number in scientific notation.
3. Change a number from scientific notation to one without exponents.

The set of numbers used as exponents in our discussion thus far has been the set of positive integers. This is the only set that can be used when exponents are defined as they were in Chapter 1. In this section, however, we would like to expand this set to include all integers (positive, negative, and zero) as exponents. This will require further definitions.

DEFINITION

If $x \neq 0$, then $x^0 = 1$. (Any number, except zero, raised to the zero power equals 1.)

To illustrate why a zero exponent is defined this way, consider the product $x^0 \cdot x^5$, if $x \neq 0$.

$$x^0 \cdot x^5 = x^{0+5} = x^5$$

Since we know that the product of x^5 and a number will be x^5 only if that number is 1, we define $x^0 = 1$.

Example 1 $5^0 = 1$ ∎

Example 2 $10^0 = 1$ ∎

Example 3 $(xyz)^0 = 1$, if $x, y, z \neq 0$ ∎

Example 4 $\left(\dfrac{5x^2}{2y^3}\right)^0 = 1$, if $x, y \neq 0$ ∎

Another definition becomes necessary if we are to allow negative numbers to be used as exponents.

DEFINITION

$$x^{-1} = \frac{1}{x}, \text{ if } x \neq 0$$

(An exponent of −1 means to find the reciprocal of the base.)

To illustrate the above definition, consider the product $x^{-1} \cdot x^{1}$, if $x \neq 0$.

$$x^{-1} \cdot x^{1} = x^{-1+1} = x^{0} = 1$$

Since we know that the product of two numbers is 1 when the numbers are reciprocals, we define $x^{-1} = \dfrac{1}{x}$, if $x \neq 0$.

We would like to also be able to use other negative numbers as exponents. We can do so if we use the above definition along with the third law of exponents.

$$\left(x^{-1}\right)^{a} = x^{-a}$$

RULE

$$x^{-a} = \left(\dfrac{1}{x}\right)^{a} = \dfrac{1}{x^{a}}, \text{ if } x \neq 0 \text{ and } a > 0$$

Example 5 $5^{-1} = \dfrac{1}{5}$ ■

Example 6 $3^{-2} = \dfrac{1}{3^{2}} = \dfrac{1}{9}$ ■

Example 7 $x^{-2} = \dfrac{1}{x^{2}}$ ■

Example 8   Simplify: $x^{-2} y^{-3}$

Solution
$$\dfrac{1}{x^{2}} \cdot \dfrac{1}{y^{3}}$$

$$\dfrac{1}{x^{2} y^{3}} \quad ■$$

Example 9   Simplify: $\dfrac{1}{x^{-2}}$

Solution
$$\dfrac{1}{\dfrac{1}{x^{2}}} = 1 \cdot \dfrac{x^{2}}{1} = x^{2} \quad ■$$

When the base of an exponential expression is a fraction and the exponent is negative, eliminate the negative exponent by finding the reciprocal of the fraction. For instance, $\left(\dfrac{x}{2}\right)^{-1}$ is $\dfrac{2}{x}$.

Example 10   Simplify: $\left(\dfrac{a}{4}\right)^{-3}$

Solution $\left(\dfrac{4}{a}\right)^{3} = \dfrac{64}{a^{3}}$ ∎

CAUTION   The negative sign of the exponent *never* affects the sign of the expression. The expression -3^{-2} is simplified as $-\dfrac{1}{3^{2}} = -\dfrac{1}{9}$, NOT $-3^{-2} = 9$.

EXERCISE 6-2-1

Simplify. Write final answers without negative exponents. Assume all variables represent nonzero numbers.

1. 3^{0}

2. $-4x^{0}$

3. $\left(\dfrac{-2}{x^{2}y^{3}}\right)^{0}$

4. x^{-3}

5. a^{-5}

6. $(ab)^{-1}$

7. 3^{-3}

8. $(-2)^{-3}$

9. $(2x)^{-3}$

10. $3^{-3}z^{-5}$

11. $\dfrac{1}{6^{-2}}$

12. $\dfrac{1}{x^{-3}}$

13. $\dfrac{1}{a^{-5}}$

14. $\dfrac{2}{y^{-5}}$

15. $\left(\dfrac{3}{4}\right)^{-2}$

16. $\left(\dfrac{1}{a^{2}}\right)^{-2}$

17. $\left(\dfrac{2}{x^{3}}\right)^{-2}$

18. $\left(\dfrac{3}{y^{2}}\right)^{-3}$

Example 19   Simplify: $\dfrac{a^2 a^{-8}}{a^{-4}}$

Solution   The numerator of the expression is a product with like bases, so we add the exponents.

$$\frac{a^{2+(-8)}}{a^{-4}} = \frac{a^{-6}}{a^{-4}}$$

Now we are dividing expressions with like bases, so we subtract the exponents.

$$a^{-6-(-4)} = a^{-2}$$

Using the definition of negative exponents, we write

$$a^{-2} = \frac{1}{a^2}. \quad \blacksquare$$

Note that when a *factor* is moved from numerator to denominator or from denominator to numerator, the sign of the exponent is changed. Thus Example 19 could be worked as follows.

$$\frac{a^2 a^{-8}}{a^{-4}} = \frac{a^2 a^4}{a^8} = \frac{a^6}{a^8} = \frac{1}{a^2}$$

This shortcut can only be used with *factors*, never with *terms*.

EXERCISE 6-2-2

Simplify. Write final answers without negative exponents. Assume all variables represent nonzero numbers.

1. $x^{-1}x^{-3}$ **2.** $a^{-3}a^{-4}$ **3.** $y^{-4}y^{-2}$ **4.** $2^{-3} \cdot 2^4$

5. $3^{-2} \cdot 3^6$ **6.** $\dfrac{x^3}{x^{-2}}$ **7.** $\dfrac{y^2}{y^{-5}}$ **8.** $\dfrac{7^5}{7^{-1}}$

9. $\dfrac{2^{-3}}{2^2}$ **10.** $\dfrac{-10a^{-1}}{a^{-4}}$ **11.** $-4z^{-1}$ **12.** $3a^{-2}$

13. $\left(x^{-2}\right)^{-4}$ **14.** $\left(a^5\right)^{-2}$ **15.** $\dfrac{x^{-2}x^{-6}}{x^7}$ **16.** $\left(x^2 y^{-3}\right)^{-2}$

17. $\left(2x^2\right)^{-3}$

18. $\dfrac{5^3 \cdot 5^0}{5^2}$

19. $\dfrac{2^{-5} \cdot 2^0}{2^{-2}}$

20. $\dfrac{-5x^{-2}y^2}{2x^{-5}y^{-1}}$

21. $\dfrac{x^3 y^{-7}}{x^3 y^0}$

22. $\left(\dfrac{x^{-1} \cdot x^0}{x^5}\right)^{-2}$

23. $\left(\dfrac{y}{z}\right)^{-4}$

24. $\left(\dfrac{3x}{2z}\right)^{-2}$

We will simplify the expressions in the remaining examples in just one way. There may be other correct ways to achieve the simplified final form.

Example 20   Simplify: $\left(-3a^2\right)^{-4}\left(b^{-1}\right)^{-3}$

Solution

$$\left(\frac{1}{-3a^2}\right)^4\left(b^3\right)$$

$$\left(\frac{1}{(-3)^4\left(a^2\right)^4}\right)\left(b^3\right)$$

$$\frac{b^3}{81a^8} \quad\blacksquare$$

Example 21   Simplify: $\dfrac{y^{-1}-x^{-1}}{x-y}$

Solution   Using the definition of negative exponents, we write

$$\frac{\dfrac{1}{y}-\dfrac{1}{x}}{x-y}.$$

Here we have a complex fraction. To simplify it we will use Method 1 from Section 5–6 and multiply both the numerator and denominator by xy (the LCD of the fractions contained in the complex fraction).

$$\frac{xy\left(\dfrac{1}{y}-\dfrac{1}{x}\right)}{xy(x-y)}$$

$$\frac{x-y}{xy(x-y)}=\frac{1}{xy} \quad\blacksquare$$

CAUTION   The expression $\dfrac{y^{-1} - x^{-1}}{x - y}$ cannot be rewritten using the shortcut described after Example 19 because y^{-1} and x^{-1} are *terms*, not *factors*.

EXERCISE 6-2-3

Simplify. Write final answers without negative exponents. Assume all variables represent nonzero numbers.

1. $\left(-2x^3\right)^{-3}\left(3x^{-1}\right)^2$

2. $\left(2x^2y^{-3}\right)^{-3}\left(-2x^{-3}y^2\right)^2$

3. $\left(2x^2y\right)^3\left(2x^2y\right)^{-2}$

4. $\left(\dfrac{x^2y^{-1}}{x^{-5}}\right)\left(\dfrac{x^{-1}y^3}{x^5}\right)^{-3}$

5. $\left(\dfrac{2^{-3}}{x^4}\right)^{-2}\left(\dfrac{-2x^3}{x^{-1}}\right)^3$

6. $\left(\dfrac{-10}{x^5}\right)^{-1}\left(\dfrac{x^{-4}}{5}\right)^{-2}\left(\dfrac{2}{5x^2}\right)$

7. $(ab)^{-1}$

8. $(a+b)^{-1}$

9. $a^{-1}+b^{-1}$

10. $\dfrac{x^{-1}+y^{-1}}{xy}$

11. $\left(x^{-1}+2^{-2}\right)^{-1}$

12. $\left(x^{-1}+y^{-1}\right)^{-2}$

Exponents are used in many fields of science to write numbers in what is called *scientific notation*. If a number is either very large or very small, this method of expressing the number keeps it from being cumbersome and can make computations easier. Calculators utilize scientific notation when numbers are too large or too small for the display.

> ### SCIENTIFIC NOTATION
> A number is in **scientific notation** if it is expressed in the form
> $$a \times 10^n$$
> where $1 \le a < 10$ and n is an integer.

For example, the number 3600 can be expressed in scientific notation as 3.6×10^3.

Example 22   The earth is approximately 93,000,000 miles from the sun. Express this number in scientific notation.

Solution $93,000,000 = 9.3 \times 10^7$ ∎

Notice that the definition is very explicit. The product must be of a number equal to or greater than 1 and less than 10, and a power of 10.   The number 93,000,000 can be expressed as 0.93×10^8. However, this is *not* scientific notation because 0.93 is not equal to or greater than 1.

Example 23   Write 2.5×10^6 without using exponents.

Solution $2.5 \times 10^6 = 2,500,000$

Note that multiplying by 10^6 moves the decimal point six places to the right. ∎

Example 24   Write 2.5×10^{-6} without using exponents.

Solution $2.5 \times 10^{-6} = 0.0000025$

Note that multiplying by 10^{-6} moves the decimal point six places to the left. ∎

Example 25   Write 0.0000000345 in scientific notation.

Solution   We see that part of the answer must be 3.45 (equal to or greater than 1 and less than 10 always gives one nonzero digit to the left of the decimal point). If we now look at 3.45 we must ask, "What power of 10 will return the decimal point to its original position?" Counting, we get eight places to the left.

$$0.0000000345 = 3.45 \times 10^{-8}$$ ∎

EXERCISE 6-2-4

State whether or not the given number is in scientific notation.

1. 3.6×10^5 **2.** 0.5×10^8 **3.** 2.78×10^{16}

4. 54.1×10^6 **5.** 8.2×10^{-3} **6.** 7.8×10^{-4}

7. 25×10^{-3} **8.** 0.645×10^4 **9.** 5×10^8 **10.** 10×10^7

Write each number in scientific notation.

11. 5000   **12.** 346,000,000   **13.** 0.000000235

14. 23,000,000,000,000   **15.** 0.0000000052   **16.** 5280

17. 68   **18.** 728   **19.** 728,000

20. 0.0000728

Write each number without using exponents.

21. 3.201×10^5 **22.** 6.23×10^8 **23.** 7.28×10^{-6}

24. 1.07×10^{-9} **25.** 5.02×10^{10} **26.** 3.58×10^{-1}

27. 4.07×10^{-5} **28.** 5.3762×10^2 **29.** 3.6×10^0

30. 9.9×10^1

31. Mars is approximately 49,000,000 miles from Earth. Express this distance in scientific notation.

32. An enormous cloud of hydrogen gas eight million miles in diameter was discovered around the comet Bennet by NASA in 1970. Express this distance in scientific notation.

33. A light year (the distance light travels in a year) is approximately 6.0×10^{13} miles. Express this distance without exponents.

34. An angstrom is a unit of length. One angstrom is approximately 1×10^{-8} cm. Express this length without exponents.

35. The diameter of the nucleus of an average atom is approximately 3.5×10^{-12} cm. Express this distance without exponents.

36. A red blood cell is approximately 0.001 cm in diameter. Express this length in scientific notation.

37. The Rubik's Cube puzzle has more than 43,000,000,000,000,000,000 color combinations. Express this number in scientific notation.

38. The Milky Way galaxy has a diameter of approximately 590,000,000,000,000,000 miles. Express this distance in scientific notation.

6–3   RADICALS

> **OBJECTIVES**
>
> Upon completion of this section you should be able to:
> 1. Give the principal root of a number.
> 2. Determine if a radical is rational or irrational.
> 3. Approximate an irrational radical.

Suppose we cube a number and the answer is 8. We want to determine what number was cubed. One way to write this is $x^3 = 8$. What is x? The answer is 2, because $2^3 = 8$. The number 2 is called the *cube root* of 8.

> **DEFINITION**
>
> a is an ***n*th root** of b if $a^n = b$.

Example 1   5 is a square root of 25 since $5^2 = 25$.   ∎

Example 2 -5 is a square root of 25 since $(-5)^2 = 25$.   ∎

Example 3 -4 is a cube root of -64 since $(-4)^3 = -64$.   ∎

Example 4 -2 is a fourth root of 16 since $(-2)^4 = 16$.   ∎

Example 5 $\dfrac{1}{4}$ is a cube root of $\dfrac{1}{64}$ since $\left(\dfrac{1}{4}\right)^3 = \dfrac{1}{64}$.   ∎

A notation used to indicate an *n*th root is **radical notation**. The symbol $\sqrt[n]{a}$ is called a **radical**. The number *n* is called the **index** and is always an integer greater than 1. If the index is 2, it is usually omitted. The expression *a* is called the **radicand**.

As can be seen in Examples 1 and 2, 25 has two square roots. The radical is used to indicate the *principal root* of a number.

RULE

The **principal *n*th root** of a, $\left(\sqrt[n]{a}\right)$, is:
    zero if $a = 0$;
    positive if $a > 0$;
    negative if $a < 0$ and n is an odd integer;
    not a real number if $a < 0$ and n is an even integer.

According to this definition, $\sqrt{25} = 5$, not -5. The last part of the definition says that an even root of a negative number is not a real number. For example, $\sqrt{-4}$ is not real since we need to find a number that when squared would give a negative result. However, when we multiply two negative numbers or two positive numbers, we get a positive result. Therefore, it is not possible for a square number to be negative in the set of real numbers.

Example 6 $\sqrt[5]{0} = 0$ The principal fifth root of zero is zero. ∎

Example 7 $\sqrt[3]{8} = 2$ ∎

Example 8 $\sqrt[3]{-8} = -2$ Since the radicand is negative and the index is odd, the principal cube root of -8 is -2. ∎

Example 9 $\sqrt{16} = 4$ The radicand is positive and the index is even, so the principal square root of 16 is 4. ∎

Example 10 $\sqrt{-64}$ is not a real number because the radicand is negative and the index is even. ∎

Example 11 $-\sqrt{16} = -4$ ∎

Example 12 $\sqrt[4]{-16}$ is not a real number. ∎

Example 13 $\sqrt{\dfrac{1}{4}} = \dfrac{1}{2}$ since $\left(\dfrac{1}{2}\right)^2 = \dfrac{1}{4}$. ∎

Since a radical is telling us to find the principal root of a number, it is considered an operation in the same way as raising a number to an exponent. Finding a principal *n*th root and raising to the *n*th power are inverse operations.

RULE

If $\sqrt[n]{a}$ is a real number, then $\left(\sqrt[n]{a}\right)^n = \sqrt[n]{a^n} = a$.

Example 14 $\left(\sqrt{2}\right)^2 = 2$ ■

Example 15 $\left(\sqrt[4]{7}\right)^4 = 7$ ■

Example 16 $\sqrt[3]{5^3} = 5$ ■

Example 17 $\sqrt{9} = \sqrt{3^2} = 3$ ■

EXERCISE 6-3-1

Evaluate. If not a real number, so state.

1. $\sqrt{4}$ **2.** $\sqrt{25}$ **3.** $\sqrt[3]{27}$ **4.** $\sqrt{1}$

5. $\sqrt[3]{-1}$ **6.** $\sqrt{\dfrac{4}{9}}$ **7.** $\sqrt{\dfrac{25}{16}}$ **8.** $\sqrt[3]{\dfrac{8}{27}}$

9. $\sqrt[3]{\dfrac{27}{125}}$ **10.** $\sqrt{-16}$ **11.** $\sqrt[5]{-32}$ **12.** $\sqrt[3]{64}$

13. $\sqrt{3^2}$ **14.** $\sqrt[9]{2^9}$ **15.** $-\sqrt{5^2}$ **16.** $-\sqrt[3]{4^3}$

17. $-\sqrt{\dfrac{36}{49}}$ **18.** $\sqrt[9]{-1}$ **19.** $-\sqrt{81}$ **20.** $-\sqrt[3]{-64}$

21. $-\sqrt{36}$ **22.** $-\sqrt[6]{-64}$ **23.** $\sqrt[87]{-1}$ **24.** $\sqrt[7]{-128}$

Radicals that are irrational numbers can be approximated as follows.

Example 18   Approximate: $\sqrt{10}$

Solution   We know $\sqrt{9}=3$ and $\sqrt{16}=4$. Since $9 < 10 < 16$, $\sqrt{9} < \sqrt{10} < \sqrt{16}$ or $3 < \sqrt{10} < 4$. So $\sqrt{10}$ is between 3 and 4. ■

If a more precise approximation is needed, a calculator may be used.

EXERCISE 6-3-2

Approximate each of the following radicals and then use a calculator to evaluate each to three decimal places.

1. $\sqrt{8}$ **2.** $\sqrt{15}$ **3.** $\sqrt[3]{8.5}$

4. $\sqrt{14.6}$ **5.** $\sqrt{21}$ **6.** $\sqrt[4]{20}$

7. $\sqrt{50}$ **8.** $\sqrt{85}$ **9.** $\sqrt{110}$

10. $\sqrt{140}$ **11.** $\sqrt{905}$ **12.** $\sqrt[3]{1008}$

6–4   RATIONAL EXPONENTS

OBJECTIVES

Upon completion of this section you should be able to:
1. State the meaning of a rational exponent.
2. Change expressions containing rational exponents to radical form.
3. Change expressions containing radicals to exponential form.
4. Evaluate expressions having rational exponents.

Having discussed radicals in the previous section, we are now ready to give meaning to an exponent that is a fraction.

DEFINITION

$a^{\frac{1}{n}} = \sqrt[n]{a}$ if n is a positive integer and $\sqrt[n]{a}$ is a real number.

To illustrate this definition, consider the product $a^{\frac{1}{2}} \cdot a^{\frac{1}{2}}$. We can rewrite this as $\left(a^{\frac{1}{2}}\right)^2$. Applying the rule for exponents for a power to a power, we multiply the exponents: $\left(a^{\frac{1}{2}}\right)^2 = a^{\frac{1}{2}\cdot 2} = a^1 = a$. Since $\left(\sqrt{a}\right)^2 = a$ and $\left(a^{\frac{1}{2}}\right)^2 = a$, we define $a^{\frac{1}{2}} = \sqrt{a}$.

Example 1 $2^{\frac{1}{2}} = \sqrt{2}$ ∎

Example 2 $x^{\frac{1}{3}} = \sqrt[3]{x}$ ∎

We can extend this idea to include all rational numbers if we combine the above definition with the third law of exponents.

$$\left(\sqrt[n]{a}\right)^m = \left(a^{\frac{1}{n}}\right)^m = a^{\frac{1}{n}\cdot m} = a^{\frac{m}{n}}$$

RULE

$a^{\frac{m}{n}} = \sqrt[n]{a^m} = \left(\sqrt[n]{a}\right)^m$ if n is a positive integer, m is an integer, and $\sqrt[n]{a}$ is a real number.

Example 3   Change to radical form: $x^{\frac{2}{3}}$

Solution   The index is 3 and the exponent is 2. $\sqrt[3]{x^2}$ or $\left(\sqrt[3]{x}\right)^2$ ∎

Example 4   Change to radical form: $a^{\frac{4}{5}}$

Solution   The index is 5 and the exponent is 4. $\sqrt[5]{a^4}$ or $\left(\sqrt[5]{a}\right)^4$ ∎

Example 5   Change to radical form: $x^{\frac{-1}{3}}$

Solution   The exponent is negative, so we need to write the reciprocal. $\dfrac{1}{x^{\frac{1}{3}}} = \dfrac{1}{\sqrt[3]{x}}$ ∎

Example 6   Change to exponential form: $\sqrt[5]{y^3}$

Solution   The index is 5, so the denominator of the exponent will be 5. $y^{\frac{3}{5}}$ ∎

Example 7   Change to exponential form: $\dfrac{-1}{\sqrt{x}}$

Solution   The index is 2, so the denominator of the exponent will be 2. $\dfrac{-1}{x^{\frac{1}{2}}} = -1 \cdot x^{-\frac{1}{2}} = -x^{-\frac{1}{2}}$ ∎

EXERCISE 6-4-1

Change from exponential form to radical form.

1. $x^{\frac{1}{3}}$ 2. $x^{\frac{1}{5}}$ 3. $-a^{\frac{3}{4}}$ 4. $y^{\frac{2}{7}}$

5. $6^{\frac{1}{2}}$ 6. $5^{\frac{2}{3}}$ 7. $a^{-\frac{3}{4}}$ 8. $x^{-\frac{5}{7}}$

9. $-3^{-\frac{1}{2}}$ 10. $7^{-\frac{2}{3}}$ 11. $(2x)^{\frac{2}{3}}$ 12. $-(3y)^{-\frac{2}{3}}$

Change from radical form to exponential form.

13. $\sqrt{x}$ **14.** $\sqrt[5]{a}$ **15.** $\sqrt[5]{x^4}$ **16.** $\sqrt[3]{y^2}$

17. $\sqrt[5]{3}$ **18.** $\left(\sqrt[3]{7}\right)^2$ **19.** $-\dfrac{1}{\sqrt{x}}$ **20.** $\dfrac{1}{\left(\sqrt[5]{x}\right)^2}$

21. $\dfrac{1}{\sqrt[3]{4}}$ **22.** $-\dfrac{1}{\sqrt[4]{7}}$ **23.** $\sqrt[3]{2y}$ **24.** $\sqrt[4]{x^3 y}$

Note that in a rational exponent the denominator represents the index and the numerator represents a power. Thus an exponent can have three parts that require you to do three operations to the base.

$$a^{-\frac{m}{n}}$$

The negative sign tells you to find the reciprocal of the base.

The denominator tells you to find the nth root of the base.

The numerator tells you to use the base as a factor m times.

The order in which these operations are done does not matter.

Example 8   Evaluate: $8^{\frac{2}{3}}$

Solution   The denominator of the exponent tells us to find the cube root of 8 and the numerator tells us to square the result.

$$\left(\sqrt[3]{8}\right)^2 = 2^2 = 4 \quad \blacksquare$$

Example 9   Evaluate: $(-32)^{\frac{2}{5}}$

Solution $\left(\sqrt[5]{-32}\right)^2 = (-2)^2 = 4 \quad \blacksquare$

Example 10 Evaluate: $-32^{-\frac{2}{5}}$

Solution This exponent contains all three operations. We must find the reciprocal of the base, the fifth root and square. Here we will find the fifth root first, then the reciprocal, and finally the square.

$$-\left(\sqrt[5]{32}\right)^{-2} = -(2)^{-2} = -\frac{1}{2^2} = -\frac{1}{4} \quad \blacksquare$$

Example 11 Evaluate: $\left(\dfrac{1}{27}\right)^{-\frac{2}{3}}$

Solution $\quad (27)^{\frac{2}{3}} = \left(\sqrt[3]{27}\right)^2 = 3^2 = 9 \quad \blacksquare$

EXERCISE 6-4-2

Evaluate.

1. $4^{\frac{1}{2}}$

2. $27^{\frac{1}{3}}$

3. $9^{\frac{1}{2}}$

4. $(-27)^{\frac{2}{3}}$

5. $-27^{\frac{2}{3}}$

6. $(8)^{\frac{2}{3}}$

7. $(-8)^{\frac{2}{3}}$

8. $(8)^{\frac{-2}{3}}$

9. $\left(\dfrac{1}{8}\right)^{\frac{-2}{3}}$

10. $\left(-\dfrac{1}{8}\right)^{\frac{-2}{3}}$

11. $-\left(\dfrac{1}{8}\right)^{\frac{-2}{3}}$

12. $\left(\dfrac{64}{27}\right)^{\frac{-2}{3}}$

13. $\left(\sqrt[5]{-1}\right)^2$

14. $\left(\sqrt[4]{16}\right)^2$

15. $\sqrt[5]{(-21)^5}$

16. $\sqrt[3]{(-2)^3}$

Rational exponents follow the same laws as integer exponents. This is shown in the following examples.

Example 12   Simplify: $x^{\frac{1}{3}} \cdot x^{\frac{1}{3}}$

Solution   We are multiplying expressions with like bases, so we keep the base and add the exponents.

$$x^{\frac{1}{3}+\frac{1}{3}} = x^{\frac{2}{3}} \quad \blacksquare$$

Example 13   Simplify: $\dfrac{a^{\frac{2}{3}}}{a^{\frac{1}{2}}}$

Solution   We are dividing expressions with like bases, so we keep the base and subtract the exponents.

$$a^{\frac{2}{3}-\frac{1}{2}} = a^{\frac{4}{6}-\frac{3}{6}} = a^{\frac{1}{6}} \quad \blacksquare$$

Example 14   Simplify: $\left(y^{\frac{2}{3}}\right)^{3} \left(25y^4\right)^{\frac{1}{2}}$

Solution   We are raising a power to a power in the first factor and factors to a power in the second factor.

$$\left(y^{\frac{2}{3}(3)}\right)\left(25^{\frac{1}{2}} y^{4\left(\frac{1}{2}\right)}\right) = y^2\left(5y^2\right) = 5y^4 \quad \blacksquare$$

Example 15   Simplify: $\sqrt{\sqrt[4]{x}}$

Solution   Rewrite using exponential form.

$$\sqrt{(x)^{\frac{1}{4}}} = \left((x)^{\frac{1}{4}}\right)^{\frac{1}{2}} = x^{\frac{1}{4}\left(\frac{1}{2}\right)} = x^{\frac{1}{8}} = \sqrt[8]{x} \quad \blacksquare$$

Example 16   Simplify: $\dfrac{\sqrt{a}}{\sqrt[3]{a}}$

Solution   Rewrite using exponential form.

$$\frac{a^{\frac{1}{2}}}{a^{\frac{1}{3}}} = a^{\frac{1}{2}-\frac{1}{3}} = a^{\frac{3}{6}-\frac{2}{6}} = a^{\frac{1}{6}} = \sqrt[6]{a} \quad \blacksquare$$

EXERCISE 6-4-3

Simplify. Write your answer without negative exponents.

1. $x^{\frac{1}{5}} \cdot x^{\frac{2}{5}}$

2. $y^{-\frac{2}{7}} \cdot y^{\frac{4}{7}}$

3. $x^{\frac{1}{3}} \cdot x^{-\frac{1}{4}}$

4. $a^{\frac{3}{5}} \cdot a^{\frac{1}{3}}$

5. $2^{\frac{2}{3}} \cdot 2^{\frac{1}{5}}$

6. $5^{\frac{3}{2}} \cdot 5^{-\frac{2}{3}}$

7. $\dfrac{10^{\frac{2}{3}}}{10^{\frac{1}{4}}}$

8. $\dfrac{a^{\frac{2}{5}}}{a^{\frac{1}{2}}}$

9. $\dfrac{x^{\frac{1}{4}}}{x^{-\frac{2}{3}}}$

10. $\dfrac{x^{\frac{1}{5}}}{x^{-\frac{1}{5}}}$

11. $\dfrac{a^{\frac{2}{3}}}{a^{-\frac{2}{5}}}$

12. $\left(y^{\frac{2}{5}}\right)^{10} \left(36y^6\right)^{\frac{1}{2}}$

13. $\left(27x^6\right)^{\frac{2}{3}} \left(2x^{\frac{2}{3}}\right)^3$

14. $\left(a^{\frac{1}{3}}\right)^6 \left(9a^2\right)^{\frac{-1}{2}}$

15. $\left(16x^{-2}\right)^{\frac{1}{4}} \left(x^4\right)^{\frac{-1}{2}}$

16. $\dfrac{\left(4x^2\right)^{\frac{-1}{2}} \left(x^{\frac{2}{3}}\right)}{2x^{\frac{1}{2}}}$

Simplify. Write your answer in radical form.

17. $\sqrt[3]{\sqrt{5}}$

18. $\sqrt[4]{\sqrt{7}}$

19. $\sqrt[3]{\sqrt[4]{x^5}}$

20. $\dfrac{\sqrt{10}}{\sqrt[3]{10}}$

21. $\dfrac{\sqrt[3]{a^5}}{\sqrt[4]{a^3}}$

22. $\dfrac{\sqrt[4]{y^3}}{\sqrt[5]{y^2}}$

6–5   SIMPLIFICATION OF RADICALS

We have now defined the principal nth root of a number and established two notations for it (the radical and the rational exponent). Radical notation is most often used for numerical expressions, while exponential notation is frequently used for variable expressions. In the remaining sections of this chapter we will discuss simplification and the operations of addition, subtraction, multiplication, and division for expressions containing radicals.

A radical is in **simplest form** when the following three conditions are satisfied.

1. No power of a factor is equal to or greater than the index.
2. The radicand does not contain a fraction and there are no radicals in the denominator of an expression.
3. The index and the exponent of the radicand have no factors in common.

In this section we will simplify radical expressions so that the first condition is satisfied. The remaining two conditions will be discussed in Section 6–7. We will assume all variables represent positive numbers to avoid expressions that will result in values that are not real numbers.

Since $\sqrt[n]{ab} = (ab)^{\frac{1}{n}}$, we can use the fourth law of exponents to write

$$\sqrt[n]{ab} = (ab)^{\frac{1}{n}} = a^{\frac{1}{n}}b^{\frac{1}{n}} = \sqrt[n]{a}\sqrt[n]{b}$$

RULE

$$\sqrt[n]{ab} = \sqrt[n]{a}\sqrt[n]{b}$$

We want to find and simplify any factor of the form $\sqrt[n]{a^n}$.

Example 1   Simplify: $\sqrt{48}$

Solution   Since the index is 2, we look for the largest factor of 48 that is a square.

$$\sqrt{16\cdot 3}$$

Now we use the rule and rewrite as a product of two square roots.

$$\sqrt{16}\sqrt{3}$$

$$4\sqrt{3} \quad \blacksquare$$

Example 2   Simplify: $2\sqrt{75}$

Solution

$$2\sqrt{25\cdot 3}$$
$$2\sqrt{25}\sqrt{3}$$
$$2\cdot 5\cdot\sqrt{3}$$
$$10\sqrt{3} \ \blacksquare$$

Example 3   Simplify: $\sqrt[3]{48}$

Solution   The index is 3, so we need to find the largest factor of 48 that is a cube.

$$\sqrt[3]{8}\sqrt[3]{6}$$
$$2\sqrt[3]{6} \ \blacksquare$$

Example 4   Simplify: $4\sqrt[5]{-64}$

Solution   Since $(-2)^5 = -32$, we write -64 as $-32\cdot 2$.

$$4\sqrt[5]{-32}\sqrt[5]{2}$$
$$4(-2)\sqrt[5]{2}$$
$$-8\sqrt[5]{2} \ \blacksquare$$

If variables are present in the radicand, the radical can be simplified using the above rule or by writing in exponential form.

Example 5   Simplify: $\sqrt{x^4 y^6}$

Solution   METHOD 1

$$\sqrt{\left(x^2\right)^2\left(y^3\right)^2}$$
$$\sqrt{\left(x^2\right)^2}\sqrt{\left(y^3\right)^2}$$
$$x^2 y^3$$

METHOD 2

$$\left(x^4 y^6\right)^{\frac{1}{2}}$$
$$x^{\frac{4}{2}}y^{\frac{6}{3}}$$
$$x^2 y^3 \ \blacksquare$$

Example 6   Simplify: $\sqrt{8x^3 y^4}$

Solution

$$\sqrt{4\cdot 2\cdot x^2\cdot x\cdot\left(y^2\right)^2}$$
$$\sqrt{4}\sqrt{x^2}\sqrt{\left(y^2\right)^2}\sqrt{2x}$$
$$2xy^2\sqrt{2x} \ \blacksquare$$

EXERCISE 6-5-1
Simplify.

1. $\sqrt{8}$ **2.** $\sqrt{18}$ **3.** $\sqrt{45}$ **4.** $\sqrt{50}$

5. $3\sqrt{20}$ **6.** $2\sqrt{75}$ **7.** $\sqrt[3]{16}$ **8.** $\sqrt[3]{-54}$

9. $\sqrt[3]{-16}$ **10.** $-2\sqrt{300}$ **11.** $\sqrt{80}$ **12.** $-5\sqrt[3]{81}$

13. $3\sqrt{98}$ **14.** $\sqrt[4]{48}$ **15.** $-7\sqrt[5]{-64}$ **16.** $7\sqrt[6]{a^{12}}$

17. $\sqrt[3]{x^6 y^9}$ **18.** $\sqrt{a^5}$ **19.** $\sqrt{x^4 y^5}$ **20.** $4\sqrt[3]{y^7}$

21. $2\sqrt[4]{x^9}$ **22.** $3\sqrt{8x^5}$ **23.** $\sqrt[4]{16y^9}$ **24.** $\sqrt[3]{8y^5}$

6–6   OPERATIONS WITH RADICALS

> **OBJECTIVES**
>
> Upon completion of this section you should be able to:
> 1. Combine similar radicals.
> 2. Multiply two radicals.
> 3. Multiply expressions containing radicals.

Since all radicals can be written in exponential form and we already have rules for adding, subtracting, multiplying, and dividing algebraic expressions containing exponents, it is not really necessary to have separate rules for these operations with radicals. However, since numerical expressions commonly use radical notation, we will develop such rules. In this section we will develop addition, subtraction, and multiplication rules. Division will be discussed in the next section.

> **DEFINITION**
>
> **Similar** (or like) **radicals** are radicals that have the same radicand and index.

Since radicals can be written using exponential notation, this definition is saying that similar radicals are really similar terms.

> **RULE**
>
> Only similar radicals can be added or subtracted. To add or subtract similar radicals, add the coefficients and use the result as the coefficient of the common radical.

Example 1   Simplify: $2\sqrt{x} + 3\sqrt{x}$

Solution   These are similar radicals. Add the coefficients.
$$(2+3)\sqrt{x}$$
$$5\sqrt{x} \quad \blacksquare$$

Example 2   Simplify: $4\sqrt[3]{2} + 3\sqrt[3]{2} - \sqrt[3]{2}$

Solution
$$(4+3-1)\sqrt[3]{2}$$
$$6\sqrt[3]{2} \quad \blacksquare$$

Example 3 Simplify: $3\sqrt{5}-5\sqrt{3}$

Solution The radicals are not similar, so this expression is already in simplest form. ∎

Example 4 Simplify: $3\sqrt{2}+\sqrt{50}+\sqrt{32}$

Solution Since the radicals in this expression are not similar, you might conclude that we cannot add. However, this would be incorrect because $\sqrt{50}$ and $\sqrt{32}$ are not in simplest form. We first simplify, then add.

$$3\sqrt{2}+5\sqrt{2}+4\sqrt{2}$$
$$(3+5+4)\sqrt{2}$$
$$12\sqrt{2} \ \blacksquare$$

Example 5 Simplify: $\sqrt[3]{16}-3\sqrt{8}$

Solution Although we can simplify, these radicals are not similar because they have different indices.

$$\sqrt[3]{8\cdot2}-3\sqrt{4\cdot2}$$
$$2\sqrt[3]{2}-3\cdot2\sqrt{2}$$
$$2\sqrt[3]{2}-6\sqrt{2} \ \blacksquare$$

CAUTION Be sure to simplify all radicals before trying to add or subtract.

EXERCISE 6-6-1

Simplify by combining similar radicals.

1. $\sqrt{3}+4\sqrt{3}-2\sqrt{3}$
2. $5\sqrt{2}-7\sqrt{2}+\sqrt{2}$
3. $3\sqrt{5}-2\sqrt{3}+4\sqrt{5}-\sqrt{3}$

4. $2\sqrt{6}+3\sqrt[3]{6}-\sqrt{6}+2\sqrt[3]{6}$
5. $2\sqrt{2}-3\sqrt[3]{2}-\sqrt{8}$
6. $\sqrt{8}-2\sqrt{18}+\sqrt{50}$

7. $\sqrt{28}-2\sqrt{63}+4\sqrt{7}$
8. $\sqrt[3]{16}-\sqrt[3]{54}+6\sqrt[3]{2}$
9. $2\sqrt{6}+\sqrt[3]{2}-\sqrt{54}-\sqrt[3]{16}$

10. $-4\sqrt{20}+3\sqrt{5}+\sqrt{45}$
11. $2\sqrt{12}-5\sqrt{3}+\sqrt{27}$
12. $3\sqrt{44}-7\sqrt{99}+5\sqrt{11}$

Recall that when we simplified radicals we used the rule $\sqrt[n]{ab} = \sqrt[n]{a}\sqrt[n]{b}$. To multiply radicals, we use the same rule but in the opposite direction. Therefore, the only condition necessary to multiply radicals is that the indices are the same.

RULE

$$\sqrt[n]{a}\sqrt[n]{b} = \sqrt[n]{ab}$$

Example 6   Multiply: $\sqrt{3}\sqrt{7}$

Solution $\sqrt{3\cdot 7} = \sqrt{21}$ ∎

Example 7   Multiply: $\left(2\sqrt{5}\right)\left(3\sqrt{2}\right)$

Solution $(2)(3)\sqrt{5\cdot 2} = 6\sqrt{10}$ ∎

Example 8   Multiply: $\sqrt[3]{4}\sqrt[3]{2}$

Solution $\sqrt[3]{8} = 2$ ∎

You could also change to exponential form to multiply.

Example 9   Multiply: $\sqrt[5]{x^2}\,\sqrt[5]{x^4}$

Solution   METHOD 1   METHOD 2

$\sqrt[5]{x^6} = x\sqrt[5]{x}$ $x^{\frac{2}{5}}x^{\frac{4}{5}} = x^{\frac{6}{5}}$

Notice that the expression $\sqrt[5]{x^6}$ is not in simplest form and must be rewritten as $x\sqrt[5]{x}$.

However, it is acceptable to use an improper fraction as an exponent. Therefore, $x^{\frac{6}{5}}$ is in simplest form when using exponential notation. ∎

Example 10   Multiply: $\sqrt{y}\,\sqrt[3]{y}$

Solution   The indices are not the same in these radicals, so we cannot use radical form to multiply. If we change to exponential form, we see that the laws of exponents apply.

$$y^{\frac{1}{2}}y^{\frac{1}{3}} = y^{\frac{3}{6}+\frac{2}{6}}$$

$$= y^{\frac{5}{6}}$$

$$= \sqrt[6]{y^5} \quad \blacksquare$$

EXERCISE 6-6-2

Multiply.

1. $\sqrt{2}\sqrt{5}$

2. $\sqrt[3]{3}\sqrt[3]{11}$

3. $\left(2\sqrt{7}\right)\left(3\sqrt{2}\right)$

4. $\sqrt{8}\sqrt{2}$

5. $\left(4\sqrt{a}\right)\left(2\sqrt{a^3}\right)$

6. $\left(\sqrt[3]{4}\right)\left(2\sqrt[3]{10}\right)$

7. $\sqrt[4]{x^3}\sqrt[4]{x^2}$

8. $\left(3\sqrt{a^3}\right)\left(\sqrt{18a^3}\right)$

9. $\left(4\sqrt{2x^3}\right)\left(-\sqrt{50x}\right)$

10. $\sqrt[3]{27a^2}\sqrt[3]{-16a^2}$

11. $\left(\sqrt[4]{a}\right)\left(\sqrt{a}\right)$

12. $\sqrt[5]{b^2}\sqrt[3]{b^2}$

When a radical expression contains more than one term, the distributive property is used to multiply.

Example 11   Multiply: $2\sqrt{2}\left(3\sqrt{3}+4\sqrt{5}\right)$

Solution   Here we multiply each term in the parentheses by $2\sqrt{2}$.

$$\left(2\sqrt{2}\right)\left(3\sqrt{3}\right)+\left(2\sqrt{2}\right)\left(4\sqrt{5}\right)$$
$$2\cdot3\cdot\sqrt{2}\sqrt{3}+2\cdot4\cdot\sqrt{2}\sqrt{5}$$
$$6\sqrt{6}+8\sqrt{10}\quad\blacksquare$$

Example 12   Multiply: $2\sqrt[3]{3}\left(\sqrt[3]{7}+\sqrt[3]{9}\right)$

Solution
$$2\sqrt[3]{3}\left(\sqrt[3]{7}\right)+2\sqrt[3]{3}\left(\sqrt[3]{9}\right)$$
$$2\sqrt[3]{21}+2\sqrt[3]{27}$$
$$2\sqrt[3]{21}+2\cdot3$$
$$2\sqrt[3]{21}+6\quad\text{or}\quad6+2\sqrt[3]{21}\quad\blacksquare$$

Note that the 6 and 2 cannot be combined.

EXERCISE 6-6-3

Multiply.

1. $\sqrt{2}\left(\sqrt{3}+\sqrt{5}\right)$

2. $\sqrt{3}\left(2\sqrt{2}-\sqrt{7}\right)$

3. $3\sqrt{3}\left(2\sqrt{5}+3\sqrt{3}\right)$

4. $\sqrt{2}\left(3\sqrt{3}+2\sqrt{2}\right)$

5. $5\sqrt{3}\left(2\sqrt{6}+\sqrt{15}\right)$

6. $\sqrt[3]{2}\left(4\sqrt[3]{4}-2\sqrt[3]{32}\right)$

7. $4\sqrt{3}\left(\sqrt{6}-\sqrt{3}+\sqrt{18}\right)$

8. $2\sqrt{2x}\left(5\sqrt{2x}-\sqrt{6x^3}\right)$

9. $\sqrt{3}\left(\sqrt{2}+\sqrt{7}\right)+\sqrt{2}\left(\sqrt{5}-\sqrt{3}\right)$

10. $\sqrt{2}\left(\sqrt{6}+\sqrt{2}\right)-\sqrt{3}\left(\sqrt{6}+\sqrt{3}\right)$

11. $\sqrt[3]{3}\left(2\sqrt[3]{9}+\sqrt[3]{2}\right)+3\sqrt[3]{2}\left(\sqrt[3]{3}-3\sqrt[3]{4}\right)$

12. $2\sqrt[3]{5}\left(\sqrt[3]{50}-\sqrt[3]{2}\right)-\sqrt[3]{2}\left(2\sqrt[3]{5}-\sqrt[3]{4}\right)$

The special patterns we learned for multiplying binomials can be applied to multiplying two-term radical expressions.

Example 13   Expand: $\left(3+4\sqrt{2}\right)\left(2-\sqrt{2}\right)$

Solution   The FOIL pattern for multiplying binomials can be used here.

$$3\cdot 2-3\sqrt{2}+\left(4\sqrt{2}\right)(2)-\left(4\sqrt{2}\right)\left(\sqrt{2}\right)$$
$$6-3\sqrt{2}+8\sqrt{2}-4\cdot 2$$
$$6+5\sqrt{2}-8$$
$$-2+5\sqrt{2}\quad\blacksquare$$

Example 14   Expand: $\left(3\sqrt{2}+5\sqrt{3}\right)\left(3\sqrt{2}-5\sqrt{3}\right)$

Solution   This multiplication problem fits the pattern $(a+b)(a-b)=a^2-b^2$.

$$\left(3\sqrt{2}\right)^2-\left(5\sqrt{3}\right)^2$$

$$9(2)-25(3)$$

$$-57 \ \blacksquare$$

Example 15   Expand: $\left(4\sqrt{5}+\sqrt{6}\right)^2$

Solution   Here we are squaring an expression with 2 terms.
This fits the pattern $(x+y)^2=(x+y)(x+y)=x^2+2xy+y^2$.

$$\left(4\sqrt{5}+\sqrt{6}\right)\left(4\sqrt{5}+\sqrt{6}\right)$$

$$\left(4\sqrt{5}\right)^2+2\left(4\sqrt{5}\right)\left(\sqrt{6}\right)+\left(\sqrt{6}\right)^2$$

$$80+8\sqrt{30}+6$$

$$86+8\sqrt{30} \ \blacksquare$$

EXERCISE 6-6-4

Expand.

1. $\left(\sqrt{2}+\sqrt{5}\right)\left(\sqrt{3}+\sqrt{2}\right)$ **2.** $\left(6+\sqrt{2}\right)\left(3-\sqrt{2}\right)$ **3.** $\left(2\sqrt{3}-\sqrt{5}\right)\left(3\sqrt{5}-\sqrt{2}\right)$

4. $\left(3\sqrt{6}+2\sqrt{3}\right)\left(2\sqrt{3}-\sqrt{2}\right)$ **5.** $\left(\sqrt{3}+5\right)\left(\sqrt{3}-5\right)$ **6.** $\left(2\sqrt{2}+\sqrt{6}\right)\left(2\sqrt{2}-\sqrt{6}\right)$

7. $\left(2\sqrt{11}-\sqrt{6}\right)^2$ **8.** $\left(2\sqrt{3}+\sqrt{2}\right)\left(2\sqrt{3}-5\sqrt{2}\right)$ **9.** $\left(\sqrt{2}+3\sqrt{5}\right)^2$

10. $\left(3\sqrt{5}-4\right)^2$ **11.** $\left(2\sqrt{3}-5\sqrt{7}\right)^2$ **12.** $\left(\sqrt[3]{2}-2\sqrt[3]{3}\right)\left(\sqrt[3]{2}+3\sqrt[3]{6}\right)$

6–7   RATIONALIZING DENOMINATORS

OBJECTIVES

Upon completion of this section you should be able to:
1. Simplify fractions having a single radical term in the denominator.
2. Simplify fractions having a denominator composed of two terms, at least one of which is a radical.

So far we have studied how to add, subtract, and multiply radical expressions. Now we will turn our attention to division. Recall that division is defined as multiplying by the reciprocal of the divisor, so we write $\sqrt[n]{a} \div \sqrt[n]{b}$ as $\sqrt[n]{a} \cdot \dfrac{1}{\sqrt[n]{b}} = \dfrac{\sqrt[n]{a}}{\sqrt[n]{b}}$.

Since $\sqrt[n]{a} = a^{\frac{1}{n}}$, we can use the fifth law of exponents to write $\dfrac{\sqrt[n]{a}}{\sqrt[n]{b}} = \dfrac{a^{\frac{1}{n}}}{b^{\frac{1}{n}}} = \left(\dfrac{a}{b}\right)^{\frac{1}{n}} = \sqrt[n]{\dfrac{a}{b}}$.

Note that, as in multiplication, the indices must be the same.

RULE

$$\frac{\sqrt[n]{a}}{\sqrt[n]{b}} = \sqrt[n]{\frac{a}{b}}, \quad b \neq 0$$

Example 1   Divide: $\sqrt{6} \div \sqrt{2}$

Solution $\dfrac{\sqrt{6}}{\sqrt{2}} = \sqrt{\dfrac{6}{2}} = \sqrt{3}$ ∎

Example 2   Divide: $\sqrt[3]{16} \div \sqrt[3]{2}$

Solution $\dfrac{\sqrt[3]{16}}{\sqrt[3]{2}} = \sqrt[3]{\dfrac{16}{2}} = \sqrt[3]{8} = 2$ ∎

This rule can also be used to simplify expressions if we write it as $\sqrt[n]{\dfrac{a}{b}} = \dfrac{\sqrt[n]{a}}{\sqrt[n]{b}}, b \neq 0$.

Example 3   Simplify: $\sqrt{\dfrac{1}{4}}$

Solution $\dfrac{\sqrt{1}}{\sqrt{4}} = \dfrac{1}{2}$ ∎

Example 4   Divide: $\sqrt{5} \div \sqrt{10}$

Solution $\dfrac{\sqrt{5}}{\sqrt{10}} = \sqrt{\dfrac{5}{10}} = \sqrt{\dfrac{1}{2}} = \dfrac{\sqrt{1}}{\sqrt{2}} = \dfrac{1}{\sqrt{2}}$ ■

In Section 6-5 we listed three conditions necessary for simplest radical form. The second condition is that the radicand does not contain a fraction and there are no radicals in the denominator of an expression. We can see that the result in Example 4 is *not* in simplest form since $\sqrt{2}$ is in the denominator. We need to remove the radical from the denominator. To accomplish this, we use the fundamental principle of fractions and multiply by a carefully chosen form of 1. This process is referred to as **rationalizing the denominator**.

Problems of rationalizing the denominator fall into two categories:

1. those that have only one term in the denominator, and
2. those that have two or more terms in the denominator.

Two different techniques are used, so we will give examples of each. Examples 5 through 8 show simplification of one-term denominators.

Example 5   Simplify: $\dfrac{1}{\sqrt{2}}$

Solution $\dfrac{1}{\sqrt{2}} \cdot \dfrac{\sqrt{2}}{\sqrt{2}} = \dfrac{\sqrt{2}}{\left(\sqrt{2}\right)^2}$

$= \dfrac{\sqrt{2}}{2}$ ■

Example 6   Simplify: $\dfrac{3}{\sqrt{12}}$

Solution   First we simplify $\sqrt{12}$.

$\dfrac{3}{\sqrt{4 \cdot 3}} = \dfrac{3}{2\sqrt{3}}$

Now we rationalize the denominator.

$\dfrac{3}{2\sqrt{3}} \cdot \dfrac{\sqrt{3}}{\sqrt{3}} = \dfrac{3\sqrt{3}}{2 \cdot 3} = \dfrac{\sqrt{3}}{2}$ ■

Example 7   Simplify: $\sqrt[3]{\dfrac{2}{5}}$

Solution $\dfrac{\sqrt[3]{2}}{\sqrt[3]{5}}$ We need to multiply the numerator and denominator by a number that gives a cube root of a cube in the denominator. We choose $\sqrt[3]{5^2}$.

$\dfrac{\sqrt[3]{2}}{\sqrt[3]{5}} \cdot \dfrac{\sqrt[3]{5^2}}{\sqrt[3]{5^2}} = \dfrac{\sqrt[3]{2}\sqrt[3]{25}}{\sqrt[3]{5^3}} = \dfrac{\sqrt[3]{50}}{5}$ ■

Example 8   Simplify: $\dfrac{a-b}{\sqrt{a+b}}$

Solution $\dfrac{a-b}{\sqrt{a+b}} = \dfrac{a-b}{\sqrt{a+b}} \cdot \dfrac{\sqrt{a+b}}{\sqrt{a+b}}$

$= \dfrac{(a-b)\sqrt{a+b}}{\sqrt{(a+b)^2}}$

$= \dfrac{(a-b)\sqrt{a+b}}{a+b}$ ■

EXERCISE 6-7-1

Simplify.

1. $\sqrt{\dfrac{4}{25}}$

2. $\sqrt{\dfrac{121}{36}}$

3. $\dfrac{\sqrt{48}}{\sqrt{8}}$

4. $\dfrac{\sqrt{20}}{\sqrt{5}}$

5. $\dfrac{\sqrt{50}}{\sqrt{2}}$

6. $\dfrac{1}{\sqrt{2}}$

7. $-\dfrac{1}{\sqrt{5}}$

8. $-\dfrac{2}{\sqrt{7}}$

9. $\dfrac{6}{\sqrt{20}}$

10. $\dfrac{1}{\sqrt[3]{2}}$

11. $\dfrac{1}{\sqrt[3]{4}}$

12. $\dfrac{-3}{\sqrt[5]{8}}$

13. $\sqrt{\dfrac{3}{5}}$

14. $\sqrt{\dfrac{1}{8}}$

15. $\dfrac{2}{\sqrt[3]{2}}$

16. $\dfrac{a}{\sqrt[3]{a}}$

17. $\dfrac{-1}{\sqrt[4]{y}}$

18. $\dfrac{1}{\sqrt[5]{9}}$

19. $\dfrac{2x}{\sqrt[5]{x^3}}$

20. $\dfrac{1}{\sqrt{x-y}}$

21. $\dfrac{1}{\sqrt{x}-3}$ $\qquad$ **22.** $\sqrt{\dfrac{x}{x+1}}$ $\qquad$ **23.** $\sqrt{\dfrac{8}{x-2}}$ $\qquad$ **24.** $\dfrac{x+y}{\sqrt{x+y}}$

When the denominator of a fraction contains two or more terms and one or more of these contains a radical, the problem of rationalizing the denominator becomes more complicated. We will discuss only the simplest type of these problems, which are problems with two terms in the denominator, one or both containing radicals with an index of two. Problems containing radicals with higher indices and more than two terms in the denominator are left for more advanced courses.

To rationalize expressions with two terms in the denominator, we need to remember the pattern $(a+b)(a-b) = a^2 - b^2$.

Example 9 Simplify: $\dfrac{3}{\sqrt{5}+1}$

Solution To rationalize the denominator in this example, we will multiply numerator and denominator by $\left(\sqrt{5}-1\right)$ since $\left(\sqrt{5}+1\right)\left(\sqrt{5}-1\right)$ fits the form $(a+b)(a-b)$.

$$\frac{3}{\sqrt{5}+1} = \frac{3}{\left(\sqrt{5}+1\right)} \cdot \frac{\left(\sqrt{5}-1\right)}{\left(\sqrt{5}-1\right)}$$

$$= \frac{3\left(\sqrt{5}-1\right)}{\sqrt{25}-1}$$

$$= \frac{3\sqrt{5}-3}{4}. \blacksquare$$

The expressions $\left(a\sqrt{b}+c\sqrt{d}\right)$ and $\left(a\sqrt{b}-c\sqrt{d}\right)$ are sometimes referred to as **conjugates**. Their product will always be a rational number. Therefore, the denominator of the resulting fraction will not contain a radical.

> **RULE**
>
> To simplify an algebraic expression in which the denominator is of the form $\left(a\sqrt{b}+c\sqrt{d}\right)$ or $\left(a\sqrt{b}-c\sqrt{d}\right)$, where $\sqrt{b}$ and $\sqrt{d}$ are not both rational, multiply the numerator and denominator by the conjugate of the denominator.

Example 10   Simplify: $\dfrac{5}{\sqrt{2}-\sqrt{3}}$

Solution

$$\frac{5}{\sqrt{2}-\sqrt{3}}=\frac{5}{\left(\sqrt{2}-\sqrt{3}\right)}\cdot\frac{\left(\sqrt{2}+\sqrt{3}\right)}{\left(\sqrt{2}+\sqrt{3}\right)}$$

$$=\frac{5\left(\sqrt{2}+\sqrt{3}\right)}{2-3}$$

$$=\frac{5\left(\sqrt{2}+\sqrt{3}\right)}{-1}$$

$$=-5\left(\sqrt{2}+\sqrt{3}\right)\quad\text{or}\quad-5\sqrt{3}-5\sqrt{2}\quad\blacksquare$$

Example 11   Simplify: $\dfrac{2}{\sqrt{5}-1}$

Solution

$$\frac{2}{\sqrt{5}-1}=\frac{2}{\left(\sqrt{5}-1\right)}\cdot\frac{\left(\sqrt{5}+1\right)}{\left(\sqrt{5}+1\right)}$$

$$=\frac{2\left(\sqrt{5}+1\right)}{5-1}$$

$$=\frac{2\left(\sqrt{5}+1\right)}{4}$$

$$=\frac{\sqrt{5}+1}{2}\quad\blacksquare$$

EXERCISE 6-7-2

Simplify.

1. $\dfrac{1}{\sqrt{2}+1}$
     2. $\dfrac{1}{\sqrt{3}-2}$
     3. $\dfrac{2}{\sqrt{5}+1}$
     4. $\dfrac{4}{3-\sqrt{5}}$

5. $\dfrac{\sqrt{2}}{\sqrt{6}-2}$
     6. $\dfrac{3}{\sqrt{2}+\sqrt{3}}$
     7. $\dfrac{1}{\sqrt{3}-\sqrt{5}}$
     8. $\dfrac{\sqrt{x}}{\sqrt{x}-1}$

9. $\dfrac{6}{\sqrt{x}+y}$

10. $\dfrac{a-1}{\sqrt{a}-1}$

11. $\dfrac{5}{\sqrt{7}-\sqrt{2}}$

12. $\dfrac{x}{\sqrt{x}+\sqrt{y}}$

Recall that there are three conditions necessary for a radical expression to be in simplest form. The first was discussed in Section 6-5 and the second in this section. The third condition is that the index and the exponent of the radicand have no factors in common.

Example 12   Simplify: $\sqrt[4]{x^2}$

Solution   The index, 4, and the exponent, 2, have a common factor of two.

$$x^{\frac{2}{4}} = x^{\frac{1}{2}} = \sqrt{x} \ \blacksquare$$

Example 13   Simplify: $\sqrt[6]{a^3b^3}$

Solution $a^{\frac{3}{6}}b^{\frac{3}{6}} = a^{\frac{1}{2}}b^{\frac{1}{2}} = (ab)^{\frac{1}{2}} = \sqrt{ab} \ \blacksquare$

EXERCISE 6-7-3
Simplify.

1. $\sqrt[6]{x^4}$

2. $\sqrt[8]{y^6}$

3. $\sqrt[3]{x^6y^9}$

4. $\sqrt[4]{16a^4b^2}$

5. $\sqrt[8]{a^6b^2}$

6. $\sqrt[6]{16x^4y^2}$

Chapter 6 Summary

The number in brackets refers to the section of the chapter that discusses the concept.

Terminology

- A number is in **scientific notation** if it is expressed in the form $a \times 10^n$, where $1 \le a < 10$ and n is an integer. [6–2]
- The symbol $\sqrt[n]{a}$ is a **radical**. [6–3]
- In the expression $\sqrt[n]{a}$, a is the **radicand** and n is the **index**. [6–3]
- If $a^n = b$, then a is an **nth root** of b. [6–3]
- The **principal nth root** of a $\left(\sqrt[n]{a}\right)$ is
  a. zero if $a = 0$,
  b. positive if $a > 0$,
  c. negative if $a < 0$ and n is an odd integer,
  d. not a real number if $a < 0$ and n is an even integer. [6–3]
- A radical is in **simplest form** when the following three conditions are satisfied. [6–5]
  1. No power of a factor is equal to or greater than the index.
  2. The radicand does not contain a fraction and there are no radicals in the denominator of an expression.
  3. The index and the exponent of the radicand have no factors in common.
- **Similar radicals** are radicals that have the same radicand and index. [6–6]
- The expressions $\left(a\sqrt{b} + c\sqrt{d}\right)$ and $\left(a\sqrt{b} - c\sqrt{d}\right)$ are sometimes referred to as **conjugates**. [6–7]

Rules and Procedures

Laws of Exponents [6–1]

1. $x^a x^b = x^{a+b}$

2. $\dfrac{x^a}{x^b} = x^{a-b}, \quad x \ne 0$

3. $\left(x^a\right)^b = x^{ab}$

4. $(xy)^a = x^a y^a$

5. $\left(\dfrac{x}{y}\right)^a = \dfrac{x^a}{y^a}, \quad y \ne 0$

Definitions for Exponents [6–2]

1. $x^0 = 1$, if $x \ne 0$.

2. $x^{-1} = \dfrac{1}{x}$, if $x \ne 0$.

3. $x^{-a} = \dfrac{1}{x^a} = \left(\dfrac{1}{x}\right)^a$, if $x \ne 0$.

Rational Exponents

- To change a rational exponent to radical form, use
$$x^{\frac{a}{b}} = \sqrt[b]{x^a} = \left(\sqrt[b]{x}\right)^a. \ [6\text{–}4]$$

Radicals

- Only similar radicals can be added or subtracted. [6–6]
- To multiply radicals, use
$$\sqrt[n]{a}\,\sqrt[n]{b} = \sqrt[n]{ab}. \ [6\text{–}6]$$
- To simplify an algebraic expression in which the denominator is of the form $\left(a\sqrt{b} + c\sqrt{d}\right)$ or $\left(a\sqrt{b} - c\sqrt{d}\right)$, where $\sqrt{b}$ and $\sqrt{d}$ are not both rational, multiply the numerator and denominator by the conjugate of the denominator. [6–7]

CHAPTER 6 REVIEW

Simplify.

1. $\left(2x^2 y^3\right)^5$

2. $\left(\dfrac{2a^3}{3b}\right)^3$

3. $\left[-3\left(x^2 y\right)^2\right]^3 \left[2xy^2\right]^2$

4. $t^8 t^{-2}$

5. $\dfrac{5^2}{5^{-1}}$

6. $\dfrac{x^2 + 3^0}{x^{-1}}$

7. $\left(x^3\right)^{-4}\left(-2x^{-2}\right)^{-3}$

8. $\left(\dfrac{2x}{y}\right)^{-2}\left(\dfrac{-x^3}{y^2}\right)^3$

9. $\dfrac{a^{-1} - b^{-1}}{a - b}$

10. $\dfrac{y^{\frac{2}{3}}}{y^{\frac{1}{2}}}$

11. $\left(4x^6\right)^{\frac{1}{2}}\left(27x^{-3}\right)^{\frac{1}{3}}$

12. $\dfrac{\sqrt{a}}{\sqrt[3]{a}}$

13. Write 54,200,000 in scientific notation.

14. Write 3.2×10^{-5} without exponents.

15. Write $-a^{\frac{2}{3}}$ in radical form.

16. Write $\sqrt[5]{y}$ in exponential form.

Evaluate.

17. $\sqrt{16}$

18. $\sqrt[3]{-125}$

19. $\sqrt[5]{32}$

20. $\sqrt[3]{8^2}$

21. $\dfrac{\sqrt{24}}{\sqrt{6}}$

22. $\left(\dfrac{4}{9}\right)^{-\frac{1}{2}}$

23. $27^{\frac{2}{3}}$

24. $-64^{-\frac{2}{3}}$

Simplify.

25. $\sqrt{50}$

26. $-3\sqrt[3]{32}$

27. $\sqrt{75a^4}$

28. $\sqrt[3]{-40x^3y^7}$

29. $3\sqrt{7} - 2\sqrt{7} + 8\sqrt{7}$

30. $\sqrt{75} + 3\sqrt{24} - 4\sqrt{27}$

31. $\left(-5\sqrt{5}\right)\left(3\sqrt{20}\right)$

32. $\left(2\sqrt{12x^3}\right)\left(3\sqrt{25x^5}\right)$

33. $\sqrt{5}\left(3\sqrt{20} - 2\sqrt{3}\right)$

34. $\left(2\sqrt{6} - \sqrt{2}\right)^2$

35. $\dfrac{2}{\sqrt{18}}$

36. $\dfrac{3}{\sqrt[3]{4}}$

37. $\dfrac{a-b}{\sqrt{a-b}}$

38. $\dfrac{-2}{\sqrt{6} - \sqrt{2}}$

39. $\sqrt[8]{a^6}$

CHAPTER 6 PRACTICE TEST

Evaluate.

1. $\sqrt[3]{-216}$

1. _____

2. $\sqrt[3]{(125)^2}$

2. _____

3. $8^{\frac{2}{3}}$

3. _____

4. $\left(\dfrac{49}{25}\right)^{-\frac{1}{2}}$

4. _____

5. Write 0.00000561 in scientific notation.

5. _____

6. Write 3.8×10^8 without exponents.

6. _____

Simplify.

7. $\left[\left(x^2 y^3\right)^4\right]^3$

7. _____

8. $\left(\dfrac{2x}{y^2}\right)^3 \left(\dfrac{y}{2x^2}\right)^5$

8. _____

9. $x^6 x^{-2}$

9. _____

10. $\dfrac{x^4 + 2^0}{x^{-2}}$

10. _____

11. $\left(a^2\right)^{-3} \left(a^{-5}\right)^2$

11. _____

12. $\left(2m^{\frac{1}{3}}\right)\left(-5m^{\frac{3}{4}}\right)$

12. _____

13. $\sqrt{98}$

13. _____

14. $\sqrt{18} + 5\sqrt{50} - 4\sqrt{8}$

14. _____

15. $\sqrt{7}\left(2\sqrt{28} - \sqrt{2}\right)$

15. _____

16. $\left(\sqrt{5} + \sqrt{3}\right)^2$

16. _____

17. $\dfrac{1}{\sqrt[3]{2}}$

17. _____

18. $\sqrt{\dfrac{3}{7}}$

18. _____

19. $\dfrac{5}{\sqrt{2} - \sqrt{x}}$

19. _____

20. $\sqrt[3]{-27ab^6}$

20. _____

Chapter 7

Survey

The following questions refer to material discussed in this chapter. Work as many problems as you can and check your answers with the answer key in the back of the book. The results will direct you to the sections of the chapter in which you need to work. If you answer all questions correctly, you may already have a good understanding of the material contained in this chapter.

1. Solve: $3x^2 + 11x - 4 = 0$

1. _____

2. Solve: $3x^2 = 5x$

2. _____

3. Solve: $5x^2 - 10 = 0$

3. _____

4. Solve by completing the square:
$5x^2 + 3 = 20x$

4. _____

5. Use the quadratic formula to solve:
$3x^2 + 8x = 5$

5. _____

298

6. Give the nature of the roots by evaluating the discriminant: $3x^2 - 7x + 5 = 0$

6. _____

7. Divide: $(5 + 3i) \div (1 - 2i)$

7. _____

8. Solve: $3x^4 - 8x^2 + 4 = 0$

8. _____

9. Solve: $\sqrt{x+1} = x - 5$

9. _____

10. The diagonal of a rectangle is thirteen inches. The length is two inches more than twice the width. Find the length and width of the rectangle.

10. _____

Quadratic Equations

The equations discussed in Chapter 2 were first-degree equations in one variable. A knowledge of exponents, radicals, and factoring is necessary for an understanding of equations of degree greater than one. In this chapter we will concentrate on second-degree equations.

7–1   QUADRATICS SOLVED BY FACTORING

> **OBJECTIVES**
>
> Upon completion of this section you should be able to:
> 1. Write a quadratic equation in standard form.
> 2. Solve a quadratic equation by factoring.

The degree of a polynomial equation in one variable is the highest exponent of the variable in any term. For instance, $3x^2 + x - 1 = 0$ is a second-degree equation and $x^5 + 3x^2 = 8$ is a fifth-degree equation. Equations of degree one (discussed in Chapter 2) are called **linear equations**. Equations of degree two are called **quadratic equations**. In this chapter we will discuss methods of solving quadratic equations.

Every equation of degree two can be written in standard form.

> **DEFINITION**
>
> The **standard form** of a quadratic equation is given by
> $$ax^2 + bx + c = 0$$
> where a, b, and c are real numbers and $a \neq 0$.

Example 1   Write $5x + 2 = 7 - 3x^2$ in standard form.

Solution   We want the equation in the form $ax^2 + bx + c = 0$. By subtracting 7 from both sides and adding $3x^2$ to both sides we obtain $3x^2 + 5x - 5 = 0$, which is in standard form. ■

EXERCISE 7-1-1

Write each equation in standard form.

1. $x^2 = 3x - 2$ **2.** $2x = 4 - x^2$ **3.** $6x^2 + 1 = 5x$

4. $4x^2 = 3x$ **5.** $2x^2 + 7 = 5x + 6$ **6.** $4x = 10 + 3x - 6x^2$

7. $x^2 + 5x = 2 - 4x^2$ **8.** $5x^2 - 10 + 3x = 2x + 8 + 2x^2$

We want to be able to solve higher-degree equations such as quadratic equations. While we will need to use the rules for solving equations from Chapter 2, these rules are not enough to solve for the variable. A property of real numbers that is useful in solving higher-degree equations is the **Principle of Zero Products**.

> **RULE**
>
> If $AB = 0$, then $A = 0$ or $B = 0$.

This principle states that if a product is zero, then at least one of the factors must be zero. This rule is the basis for solving higher-degree equations. If we have an equation where a polynomial equals 0 and we can factor the polynomial, then according to the Principle of Zero Products, at least one of the factors must be zero. So we set each factor equal to zero. In this way we transform a higher-degree equation into two or more simpler equations which we then solve.

Example 2   Solve: $x^2 + x - 6 = 0$

Solution   We have a trinomial on the left side of the equation and the right side equals 0. Factor.

$$(x + 3)(x - 2) = 0$$

In this form we have a product of two factors equal to zero. This means that at least one factor is zero.

Set each factor equal to zero.

$$x + 3 = 0 \quad \text{or} \quad x - 2 = 0$$

Solve the resulting equations.

$$x + 3 = 0 \quad \text{or} \quad x - 2 = 0$$
$$x = -3 \quad \text{or} \quad x = 2$$

Either of these values for x will make one of the factors zero. Therefore, the product will be zero. Since the factored form is equivalent to the original equation, these two values are solutions of the equation.

Check

$$x = -3$$
$$(-3)^2 + (-3) - 6 = 0$$
$$9 - 3 - 6 = 0$$
$$0 = 0$$

$$x = 2$$
$$(2)^2 + (2) - 6 = 0$$
$$4 + 2 - 6 = 0$$
$$0 = 0 \quad \blacksquare$$

A **solution set** consists of *all* values of the variable for which the equation is true. A solution set is indicated using braces. The solution set for Example 2 is $\{-3, 2\}$.

A PROCEDURE FOR SOLVING A QUADRATIC EQUATION BY FACTORING

1. Write the equation in standard form.
2. Factor.
3. Set each factor equal to zero and solve.

Example 3   Solve: $5x^2 - 3x - 2 = 0$

Solution

$$(5x + 2)(x - 1) = 0$$
$$5x + 2 = 0 \quad \text{or} \quad x - 1 = 0$$
$$5x = -2 \quad \text{or} \quad x = 1$$
$$x = -\frac{2}{5}$$

Both values check in the original equation. The solution set is $\left\{-\frac{2}{5}, 1\right\}$. $\blacksquare$

Example 4   Solve: $4x^2 - 2x - 6 = 0$

Solution
$$2(2x^2 - x - 3) = 0$$
$$2(2x - 3)(x + 1) = 0$$

Here we have three factors and since we know that $2 \neq 0$, then

$$2x - 3 = 0 \text{ or } x + 1 = 0.$$
$$2x = 3 \text{ or } x = -1$$
$$x = \frac{3}{2}$$

The solution set is $\left\{-1, \frac{3}{2}\right\}$. ∎

Example 5   Solve: $\dfrac{x}{2} = \dfrac{5}{2} - \dfrac{3}{x}$

Solution   This is not a quadratic equation as written. However, multiplying both sides of this equation by a common denominator will result in a quadratic equation. The LCD is $2x$.

$$2x\left(\frac{x}{2}\right) = 2x\left(\frac{5}{2} - \frac{3}{x}\right)$$

$$\not{2}x\left(\frac{x}{\not{2}}\right) = \not{2}x\left(\frac{5}{\not{2}}\right) - 2\not{x}\left(\frac{3}{\not{x}}\right)$$

$$x^2 = 5x - 6$$

$$x^2 - 5x + 6 = 0$$

We now have a quadratic equation that can be solved by factoring.

$$(x - 3)(x - 2) = 0$$
$$x - 3 = 0 \text{ or } x - 2 = 0$$
$$x = 3 \text{ or } \qquad x = 2$$

Remember, it is *necessary* to check these values in the original equation since there are variables in the denominators. Both values check in the original equation. Therefore, the solution set is $\{2, 3\}$. ∎

EXERCISE 7-1-2

Solve by factoring.

1. $x^2 + 3x + 2 = 0$ 2. $x^2 + 8x + 15 = 0$ 3. $x^2 + 8x + 7 = 0$

4. $y^2 - 4y + 3 = 0$

5. $z^2 + 2z = 8$

6. $m^2 = 3m + 10$

7. $x^2 + 18 = 9x$

8. $x^2 + 8 = 6x$

9. $(x + 5)(x - 1) = 16$

10. $(t - 6)(t + 1) = 18$

11. $5p^2 + 10p = 15$

12. $2x^2 + 5x + 3 = 0$

13. $7x^2 + 14x + 7 = 0$

14. $2x^2 - 12x + 18 = 0$

15. $3x^2 = 6x + 45$

16. $5 = 6x^2 - 13x$

17. $12 - 2x^2 = 5x$

18. $x(x + 1) = 2$

19. $y(y - 1) = 6$

20. $6 = y(y + 5)$

21. $x + 7 + \dfrac{10}{x} = 0$

22. $x = 2 + \dfrac{35}{x}$

23. $x + \dfrac{17}{6} + \dfrac{5}{3x} = 0$

24. $5x + \dfrac{4}{3x} = \dfrac{23}{3}$

7-2 INCOMPLETE QUADRATIC EQUATIONS

OBJECTIVE

Upon completion of this section you should be able to solve an incomplete quadratic equation.

When a quadratic equation is placed in standard form, $ax^2 + bx + c = 0$, and either $b = 0$ or $c = 0$, the equation is an **incomplete quadratic equation**. We will first discuss how to solve an incomplete quadratic equation when $c = 0$ (Examples 1 and 2) and then demonstrate how to solve an incomplete quadratic equation when $b = 0$ (Examples 3, 4, and 5).

The equation $5x^2 - 10x = 0$ is an incomplete quadratic equation since the third term is missing and therefore $c = 0$. When you encounter an equation of the form $ax^2 + bx = 0$, you can solve by factoring.

Example 1   Solve: $3x^2 - 2x = 0$

Solution   Factor the GCF, which is x.

$$x(3x - 2) = 0$$
$$x = 0 \text{ or } 3x - 2 = 0$$
$$3x = 2$$
$$x = \frac{2}{3}$$

The solution set is $\left\{0, \frac{2}{3}\right\}$. ∎

Notice that if the c term is missing, you can always factor x from the other terms. This means that in all such equations, zero will be one of the solutions.

Example 2   Solve: $7x^2 + 14x = 0$

Solution $7x(x + 2) = 0$
$$7x = 0 \text{ or } x + 2 = 0$$
$$x = 0 \text{ or } x = -2$$

The solution set is $\{-2, 0\}$. ∎

For an incomplete quadratic equation with the bx term missing, solution by factoring will only be possible when $ax^2 + c = 0$ results in a difference of squares. Consider $x^2 - 16 = 0$. This binomial is a difference of squares and can be factored.

$$(x+4)(x-4) = 0$$
$$x+4 = 0 \quad \text{or} \quad x-4 = 0$$
$$x = -4 \quad \text{or} \quad x = 4$$

This solution is often shortened to $x = \pm 4$.

Recall that one way of writing "What number squared is 16?" is $x^2 = 16$. Since both -4 and 4 are square roots of 16, we can write $x = \pm\sqrt{16} = \pm 4$. In general, we have the following rule.

RULE

If $A^2 = B$, then $A = \pm\sqrt{B}$.

Not all quadratic equations of the form $ax^2 + c = 0$ can be factored. If such an equation cannot be factored, we use the above rule to solve it. When we use this rule, we will say we are taking the square root of both sides of the equation.

Example 3   Solve: $x^2 - 12 = 0$

Solution   Since $x^2 - 12$ is not a difference of squares, it cannot be factored. So we solve for the x^2 term and take the square root of both sides.

$$x^2 - 12 = 0$$
$$x^2 = 12$$
$$x = \pm\sqrt{12}$$
$$x = \pm\sqrt{4 \cdot 3}$$
$$x = \pm 2\sqrt{3}$$

Notice that this gives two solutions, $+2\sqrt{3}$ and $-2\sqrt{3}$. We can write the solutions using set notation as $\left\{-2\sqrt{3}, 2\sqrt{3}\right\}$ or simply as $\left\{\pm 2\sqrt{3}\right\}$. ∎

A PROCEDURE FOR SOLVING A QUADRATIC EQUATION OF THE FORM $ax^2 + c = 0$

1. Solve for x^2.
2. Find the square root of both sides of the equation.
3. Simplify.

Example 4   Solve: $2x^2 - 10 = 0$

Solution

$$2x^2 = 10$$
$$x^2 = 5$$
$$x = \pm\sqrt{5}$$

The solution set is $\left\{\pm\sqrt{5}\right\}$. ∎

Example 5   Solve: $x^2 + 25 = 0$

Solution

$$x^2 = -25$$

There is no real solution. ∎

Note that in Example 5 we have the square of a number equal to a negative number. This can never be true in the real number system, and therefore we have no real solution.

Example 6   Solve: $(2x + 3)^2 = 5$

Solution   This equation fits the pattern $A^2 = B$. We have a squared term on the left side and a number on the right side. So we will find the square root of both sides.

$$2x + 3 = \pm\sqrt{5}$$
$$2x = -3 \pm \sqrt{5}$$
$$x = \frac{-3 \pm \sqrt{5}}{2}$$

The solution set is $\left\{\dfrac{-3-\sqrt{5}}{2}, \dfrac{-3+\sqrt{5}}{2}\right\}$. ∎

EXERCISE 7-2-1

Solve.

1. $x^2 + 3x = 0$

2. $x^2 - 5x = 0$

3. $2x^2 - 3x = 0$

4. $3y^2 + y = 0$

5. $y^2 = 8y$

6. $x^2 + x = 0$

7. $3x^2 = 5x$

8. $4m^2 + 12m = 0$

9. $5p^2 = 10p$

10. $2x^2 + 3x = 7x$

11. $4 = z^2$

12. $9 = a^2$

13. $10 = 2x^2$

14. $-26 = -2x^2$

15. $3x^2 = 60$

16. $x^2 - 32 = 0$

17. $y^2 - 14 = 0$

18. $x^2 + 16 = 0$

19. $x^2 + 16 = 12$

20. $5x^2 = 35$

21. $4b^2 = 7$

22. $3x^2 = 10$

23. $(x+1)^2 = 4$

24. $(x-3)^2 = 1$

25. $(3x+1)^2 = 5$

26. $(2x-5)^2 = 2$

27. $5m(m+1) = 3m$

28. $2t(t-3) = t$

29. $3x = \dfrac{21}{x}$

30. $5x = \dfrac{45}{x}$

7–3 SOLVING QUADRATIC EQUATIONS BY COMPLETING THE SQUARE

OBJECTIVES

Upon completion of this section you should be able to:
1. Supply the value of the constant term that will make a trinomial a perfect square trinomial.
2. Solve a quadratic equation by completing the square.

Consider the quadratic equation $x^2 - 6x + 4 = 0$. Attempting to solve this equation by the methods we now know leads to a dead end. We are not dealing with an incomplete quadratic equation, but neither can we factor the trinomial expression on the left side of the equation.

Since not all quadratic expressions are factorable, it is necessary to have another method for finding solutions. The method of *completing the square* is the next method we will discuss. It is based on the square root method and the forms

$$(x+d)^2 = x^2 + 2dx + d^2 \quad \text{and} \quad (x-d)^2 = x^2 - 2dx + d^2.$$

In the trinomial $x^2 + 2dx + d^2$ the last term is a square and it is the square of one half the coefficient of x, so $\left[\frac{1}{2}(2d)\right]^2 = d^2$. Comparing the expression $x^2 + 2dx + d^2$ to the expression $ax^2 + bx + c$, we see that for a perfect square trinomial $b = 2d$ and $c = d^2$.

Example 1   Find c such that $x^2 + 6x + c$ will be a perfect square trinomial.

Solution   Since the coefficient of x is 6, $c = \left[\frac{1}{2}(6)\right]^2 = (3)^2 = 9$.

Thus $x^2 + 6x + 9$ is a perfect square trinomial that factors into $(x+3)^2$. ∎

The process of finding the value of c so that a trinomial is a perfect square trinomial is known as **completing the square**.

Example 2   Complete the square: $x^2 - 7x + c$

Solution   The coefficient of x is –7. $c = \left[\frac{1}{2}(-7)\right]^2 = \left(-\frac{7}{2}\right)^2 = \frac{49}{4}$

$x^2 - 7x + \frac{49}{4}$

The factored form is $\left(x - \frac{7}{2}\right)^2$. ∎

EXERCISE 7-3-1

Find c in each of the following such that the resulting trinomial will be a perfect square trinomial. Then factor the perfect square trinomial.

1. $x^2 + 8x + c$ **2.** $x^2 + 10x + c$ **3.** $x^2 - 22x + c$

4. $x^2 - 4x + c$ **5.** $x^2 - 9x + c$ **6.** $x^2 - 3x + c$

7. $x^2 + 5x + c$ **8.** $x^2 + x + c$

We will now solve quadratic equations by completing the square using the following procedure.

A PROCEDURE FOR SOLVING A QUADRATIC EQUATION BY COMPLETING THE SQUARE

1. Write the equation in standard form.
2. If the coefficient of x^2 is not 1, divide each side of the equation by the coefficient of x^2.
3. Add and subtract terms so that the x^2 and x terms are on one side of the equation and the numerical term is on the other side.
4. Complete the square and add this number to both sides of the equation.
5. Factor the completed square.
6. Take the square root of both sides of the equation, remembering that if $A^2 = B$, then $A = \pm\sqrt{B}$.
7. Solve for the two values of x.

Example 3   Solve $x^2 - 15 = -2x$ by completing the square.

Solution   Write the equation in standard form. $x^2 + 2x - 15 = 0$

The coefficient of x^2 is already 1.

Add 15 to both sides. $x^2 + 2x = 15$

Complete the square.

Since $\left[\dfrac{1}{2}(2)\right]^2 = (1)^2 = 1,$ we add 1 to both sides. $x^2 + 2x + 1 = 15 + 1$

Factor. $(x+1)^2 = 16$

Find the square root of both sides and simplify. $x + 1 = \pm\sqrt{16}$

$x + 1 = \pm 4$

Solve for the two values of x. $x + 1 = 4$ or $x + 1 = -4$

$x = 3 \qquad\qquad x = -5$

The solution set is $\{-5, 3\}$. ∎

Example 4   Solve $2x^2 + 3x - 5 = 0$ by completing the square.

Solution   The equation is in standard form. $2x^2 + 3x - 5 = 0$

Divide both sides by 2. $x^2 + \dfrac{3}{2}x - \dfrac{5}{2} = 0$

Add $\dfrac{5}{2}$ to both sides. $x^2 + \dfrac{3}{2}x = \dfrac{5}{2}$

Complete the square.

Since $\left[\left(\dfrac{1}{2}\right)\left(\dfrac{3}{2}\right)\right]^2 = \left(\dfrac{3}{4}\right)^2 = \dfrac{9}{16},$

we add $\dfrac{9}{16}$ to both sides. $x^2 + \dfrac{3}{2}x + \dfrac{9}{16} = \dfrac{5}{2} + \dfrac{9}{16}$

Factor. $\left(x + \dfrac{3}{4}\right)^2 = \dfrac{49}{16}$

Find the square root of both sides and simplify.

$$x + \dfrac{3}{4} = \pm\sqrt{\dfrac{49}{16}}$$

$$x + \dfrac{3}{4} = \pm\dfrac{7}{4}$$

Solve for the two values of x. $x + \dfrac{3}{4} = \dfrac{7}{4}$ or $x + \dfrac{3}{4} = -\dfrac{7}{4}$

$$x = \dfrac{4}{4} \qquad\qquad x = -\dfrac{10}{4}$$

$$x = 1 \qquad\qquad x = -\dfrac{5}{2}$$

The solution set is $\left\{-\dfrac{5}{2}, 1\right\}$. ∎

Example 5   Solve $x^2 - 6x + 4 = 0$ by completing the square.

Solution $x^2 - 6x + 4 = 0$

$$x^2 - 6x = -4 \qquad \left[\frac{1}{2}(-6)\right]^2 = (-3)^2 = 9$$

$$x^2 - 6x + 9 = -4 + 9$$

$$(x - 3)^2 = 5$$

$$x - 3 = \pm\sqrt{5}$$

$$x - 3 = \sqrt{5} \text{ or } x - 3 = -\sqrt{5}$$

$$x = 3 + \sqrt{5} \qquad x = 3 - \sqrt{5}$$

The solution set is $\left\{3 - \sqrt{5}, 3 + \sqrt{5}\right\}$ which can also be written as $\left\{3 \pm \sqrt{5}\right\}$. ∎

Example 6   Solve $2x^2 + x + 3 = 0$ by completing the square.

Solution $$2x^2 + x + 3 = 0$$

$$x^2 + \frac{1}{2}x + \frac{3}{2} = 0$$

$$x^2 + \frac{1}{2}x = -\frac{3}{2} \qquad \left[\frac{1}{2}\left(\frac{1}{2}\right)\right]^2 = \left(\frac{1}{4}\right)^2 = \frac{1}{16}$$

$$x^2 + \frac{1}{2}x + \frac{1}{16} = -\frac{3}{2} + \frac{1}{16}$$

$$\left(x + \frac{1}{4}\right)^2 = -\frac{23}{16}$$

In this step, we see that the square of a number is $-\dfrac{23}{16}$. The square of a real number is never negative; therefore, the equation has *no real solution*. ∎

EXERCISE 7-3-2

Solve by completing the square.

1. $x^2 + 4x - 5 = 0$ **2.** $x^2 - 3 = 2x$ **3.** $x^2 - 8x + 7 = 0$

4. $x^2 + 20 = 4x$

5. $x^2 + 3x - 1 = 0$

6. $x^2 - 3 = 5x$

7. $x^2 - 2 = x$

8. $y^2 + y + 5 = 0$

9. $2x^2 + 3x - 2 = 0$

10. $2r^2 - 5r + 2 = 0$

11. $3t^2 + 6t - 4 = 0$

12. $5x^2 + 3 = 20x$

13. $2x^2 + 3x + 4 = 0$

14. $2x^2 + \dfrac{1}{2} = 2x$

15. $\dfrac{2}{3}x^2 + 2x - \dfrac{1}{2} = 0$

7–4   SOLVING QUADRATIC EQUATIONS BY FORMULA

OBJECTIVES

Upon completion of this section you should be able to:
1. State the quadratic formula.
2. Solve quadratic equations using the quadratic formula.

In a sense, a formula is the solution to all problems of a particular type. Since all quadratic equations can be put in the form $ax^2 + bx + c = 0$, the solution of this general equation in standard form will yield a formula, called the *quadratic formula*, for the solution of any quadratic equation.

We will solve the general quadratic equation by the method of completing the square.

$$ax^2 + bx + c = 0$$

Divide by a.
$$x^2 + \frac{b}{a}x + \frac{c}{a} = 0$$

Subtract $\frac{c}{a}$ from both sides.
$$x^2 + \frac{b}{a}x = -\frac{c}{a}$$

Since $\left[\left(\frac{1}{2}\right)\left(\frac{b}{a}\right)\right]^2 = \frac{b^2}{4a^2}$,

we will add $\frac{b^2}{4a^2}$ to both sides.
$$x^2 + \frac{b}{a}x + \frac{b^2}{4a^2} = \frac{b^2}{4a^2} - \frac{c}{a}$$

Factor the left side.
$$\left(x + \frac{b}{2a}\right)^2 = \frac{b^2 - 4ac}{4a^2}$$

Find the square root of both sides.
$$x + \frac{b}{2a} = \pm\sqrt{\frac{b^2 - 4ac}{4a^2}}$$

Simplify.
$$x + \frac{b}{2a} = \pm\frac{\sqrt{b^2 - 4ac}}{2a}$$

Subtract $\frac{b}{2a}$ from both sides.
$$x = -\frac{b}{2a} \pm \frac{\sqrt{b^2 - 4ac}}{2a}$$
$$x = \frac{-b \pm \sqrt{b^2 - 4ac}}{2a}$$

Thus $x = \dfrac{-b + \sqrt{b^2 - 4ac}}{2a}$ or $x = \dfrac{-b - \sqrt{b^2 - 4ac}}{2a}$.

> **RULE**
>
> If $ax^2 + bx + c = 0$ and $a \neq 0$, then $x = \dfrac{-b \pm \sqrt{b^2 - 4ac}}{2a}$. This is known as the **quadratic formula**.

> **RULE**
>
> To solve a quadratic equation using the quadratic formula:
> 1. Write the equation in standard form.
> 2. Substitute the values of a, b, and c into the formula.
> 3. Simplify.

Example 1   Use the quadratic formula to solve: $5x^2 + 3x - 8 = 0$

Solution   The equation is already in standard form, thus $a = 5$, $b = 3$, and $c = -8$.

$$x = \frac{-3 \pm \sqrt{(3)^2 - 4(5)(-8)}}{2(5)}$$

$$x = \frac{-3 \pm \sqrt{9 + 160}}{10}$$

$$x = \frac{-3 \pm \sqrt{169}}{10}$$

$$x = \frac{-3 \pm 13}{10}$$

$$x = \frac{-3 + 13}{10} \quad \text{or} \quad x = \frac{-3 - 13}{10}$$

$$x = \frac{10}{10} \qquad\qquad x = \frac{-16}{10}$$

$$x = 1 \qquad\qquad x = -\frac{8}{5}$$

The solution set is $\left\{ -\dfrac{8}{5}, 1 \right\}$. ∎

Example 2   Solve: $x^2 - 5x + 12 = 0$

Solution   Again we see that the equation is in standard form, so $a = 1$, $b = -5$, and $c = 12$.

$$x = \frac{-(-5) \pm \sqrt{(-5)^2 - 4(1)(12)}}{2(1)}$$

$$x = \frac{5 \pm \sqrt{25 - 48}}{2}$$

$$x = \frac{5 \pm \sqrt{-23}}{2}$$

Since $\sqrt{-23}$ is not a real number, the equation has *no real solution*. ∎

Example 3   Solve: $x^2 = 4(x+1)$

Solution   First we must write the equation in standard form.

$$x^2 = 4x + 4$$

$$x^2 - 4x - 4 = 0$$

$a = 1$, $b = -4$, and $c = -4$.

$$x = \frac{-(-4) \pm \sqrt{(-4)^2 - 4(1)(-4)}}{2(1)}$$

$$x = \frac{4 \pm \sqrt{16 - (-16)}}{2}$$

$$x = \frac{4 \pm \sqrt{32}}{2}$$

$$x = \frac{4 \pm 4\sqrt{2}}{2}$$

$$x = \frac{2(2 \pm 2\sqrt{2})}{2}$$

$$x = 2 \pm 2\sqrt{2}$$

The solution set is $\{2 \pm 2\sqrt{2}\}$. ∎

EXERCISE 7-4-1

Use the quadratic formula to solve. Write all solutions in simplest form.

1. $x^2 + 2x - 15 = 0$ **2.** $x^2 - 9x + 20 = 0$ **3.** $5x^2 - 7x - 6 = 0$

4. $6x^2 = x + 2$ **5.** $x^2 + 3x + 1 = 0$ **6.** $2y^2 = y + 5$

7. $2p^2 + 1 = 4p$ **8.** $4z^2 + 5z - 1 = 0$ **9.** $x(3x - 2) = 7$

10. $4 = 3x - x^2$ **11.** $3r^2 + 6r + 2 = 0$ **12.** $5x^2 + 7x + 1 = 0$

13. $2t^2 - 6t + 3 = 0$ **14.** $x(x + 2) = 7$ **15.** $x^2 = 5(2x - 5)$

16. $2x^2 - 3x + 5 = 0$ **17.** $9 + \dfrac{12}{x} + \dfrac{4}{x^2} = 0$ **18.** $3p + 8 = \dfrac{5}{p}$

19. $\dfrac{5}{6}m^2 - m + \dfrac{1}{3} = 0$ **20.** $\dfrac{3}{10}x^2 + \dfrac{2}{5}x - \dfrac{1}{2} = 0$

Practice your skills. Solve using an appropriate method.

21. $12 = 3x^2$ **22.** $3x^2 - 7x = -2$ **23.** $(p - 5)^2 = 3$

24. $x(x - 3) = 4$ **25.** $y(y + 2) + 4 = 0$ **26.** $\dfrac{x}{3} = \dfrac{x+1}{x}$

27. $6 = x(x + 1)$ **28.** $1 - \dfrac{3}{n} = \dfrac{5}{n^2}$ **29.** $\dfrac{1}{x+1} + \dfrac{1}{x-3} = \dfrac{2}{3}$

7–5   WORD PROBLEMS

> **OBJECTIVES**
>
> Upon completion of this section you should be able to:
> 1. Identify word problems that require a quadratic equation for their solution.
> 2. Solve word problems involving quadratic equations.

Certain types of word problems will lead to quadratic equations. As always, the solutions to such problems must be checked in the problem itself rather than in the resulting equation. This is necessary because the physical restrictions within the problem may eliminate one or more of the solutions. When you begin a problem, you will not know what type of equation the translation will give. The following examples should look familiar to you. They are similar to problems already presented in this text. However, each of these requires the solution of a quadratic equation.

Example 1   The length of a rectangle is one inch more than twice the width, and the area is 55 square inches. Find the length and width.

Solution   This type of problem was introduced in Section 3–2.
Let w = the width of the rectangle. Then $2w + 1$ = the length of the rectangle.

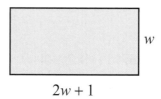

w

$2w + 1$

Since we know the area of the rectangle, we can use the formula A = length × width. Substitute the expressions for length and width in the formula to get the equation for this problem.

$$55 = (2w+1)(w)$$
$$55 = 2w^2 + w$$
$$0 = 2w^2 + w - 55$$
$$0 = (2w+11)(w-5)$$
$$2w+11 = 0 \qquad \text{or} \qquad w-5 = 0$$
$$2w = -11 \qquad\qquad\qquad w = 5$$
$$w = -\frac{11}{2}$$

At this point, we see that the solution $w = -\dfrac{11}{2}$ is not valid, since w represents a measurement of the width and negative numbers are not used for such measurements.

Therefore, the width, w, is 5 inches and the length is $(2w + 1) = 11$ inches. ■

In a right triangle, the square of the hypotenuse is equal to the sum of the squares of the two legs. This relationship is known as the Pythagorean Theorem.

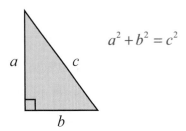

$$a^2 + b^2 = c^2$$

Example 2   The length of one leg of a right triangle is 3 cm more than the length of the other leg. If the hypotenuse is 5 cm long, what are the lengths of the legs?

Solution   Let x = the length of the shorter leg.
Then $x + 3$ = the length of the longer leg.
We use the Pythagorean Theorem to write the equation.

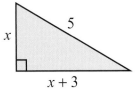

$$x^2 + (x+3)^2 = 5^2$$
$$x^2 + x^2 + 6x + 9 = 25$$
$$2x^2 + 6x - 16 = 0$$
$$2(x^2 + 3x - 8) = 0$$

We will need to use the quadratic formula to solve the equation.

$$x = \frac{-3 \pm \sqrt{(-3)^2 - 4(1)(-8)}}{2(1)}$$

$$x = \frac{-3 \pm \sqrt{9 + 32}}{2}$$

$$x = \frac{-3 \pm \sqrt{41}}{2}$$

The length of a leg must be positive, so $x = \dfrac{-3 + \sqrt{41}}{2}$. One leg is $\dfrac{-3 + \sqrt{41}}{2}$ cm and

the other leg is $x + 3 = \dfrac{-3 + \sqrt{41}}{2} + 3 = \dfrac{3 + \sqrt{41}}{2}$ cm. ■

Example 3   Greg travels upstream 105 miles and back. It takes him 4 hours longer to make the trip upstream than to come back downstream. If the speed of the river current is ten miles per hour, what is the speed of Greg's boat?

Solution   Distance-rate-time problems were introduced in Section 3–3.
Let x = the speed of the boat. We organize the given information in a table.

	r $\cdot$	t =	d
Upstream	$x - 10$		105
Downstream	$x + 10$		105

We need to write expressions for the times. The formula $rt = d$ can be written as $t = \dfrac{d}{r}$.

So we find expressions for the times by dividing the distance by the rates.

	r $\cdot$	t =	d
Upstream	$x - 10$	$\dfrac{105}{x-10}$	105
Downstream	$x + 10$	$\dfrac{105}{x+10}$	105

We write an equation that relates the times. We know that it takes 4 hours longer to go upstream than to go downstream.

$$\underbrace{\frac{105}{x-10}}_{\substack{\text{Time to go} \\ \text{upstream}}} = \underbrace{\frac{105}{x+10}+4}_{\substack{\text{Time to go} \\ \text{downstream} \quad +4}}$$

The LCD is $(x-10)(x+10)$.

$$(x-10)(x+10)\left(\frac{105}{x-10}\right) = (x-10)(x+10)\left(\frac{105}{x+10}+4\right)$$

$$105(x+10) = 105(x-10)+4(x-10)(x+10)$$

$$105x+1050 = 105x-1050+4(x^2-100)$$

$$105x+1050 = 105x-1050+4x^2-400$$

$$1050 = 4x^2-1450$$

$$2500 = 4x^2$$

$$625 = x^2$$

$$\pm 25 = x$$

The rate of the boat will not be negative, so Greg's boat travels at an average speed of 25 miles per hour. ■

Example 4   Two pipes, A and B, together fill a gasoline tank truck in two hours. Pipe A alone can fill the tank in three hours less than pipe B alone. Find the time it would take pipe A to fill the tank alone.

Solution   Work problems were discussed in Section 5–7. Recall that the equation is based on the amount of work done in one unit of time.

	Time to Fill the Tank	Amount of Tank Filled in One Hour
A	$x-3$	$\dfrac{1}{x-3}$
B	x	$\dfrac{1}{x}$
Both	2	$\dfrac{1}{2}$

$$\frac{1}{x-3}+\frac{1}{x}=\frac{1}{2}$$
$$2x(x-3)\left(\frac{1}{x-3}+\frac{1}{x}\right)=2x(x-3)\left(\frac{1}{2}\right)$$
$$2x+2(x-3)=x(x-3)$$
$$2x+2x-6=x^2-3x$$
$$0=x^2-7x+6$$
$$0=(x-1)(x-6)$$
$$x-1=0 \quad\text{or}\quad x-6=0$$
$$x=1 \qquad\qquad x=6$$

Notice that if we check the problem using $x=1$, we obtain the result that pipe B takes 1 hour and pipe A takes -2 hours. Time cannot be negative, so $x=1$ is not an option. If $x=6$, then $x-3=3$. So pipe A can fill the tank in 3 hours. ∎

Example 5   A ball is thrown downward from the observatory deck of a building with a velocity of 16 feet per second. The distance to street level is 320 feet. How long will the ball take to reach the ground? The formula that gives the height of the ball above ground is $s=-16t^2-16t+320$, where s is the distance above ground and t is the number of seconds since the ball was thrown.

Solution   The formula gives the distance above ground of the ball, so we want s to equal zero.
$$0=-16t^2-16t+320$$
$$0=-16(t^2+t-20)$$
$$0=-16(t+5)(t-4)$$
$$t+5=0 \quad\text{or}\quad t-4=0$$
$$t=-5 \qquad\qquad t=4$$

Time cannot be negative, so the time it takes for the ball to reach the ground is 4 seconds. ∎

EXERCISE 7-5-1

Solve.

1. The length of a rectangle is two feet more than twice its width. Find the dimensions of the rectangle if its area is 24 square feet.

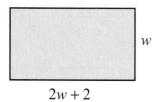

$2w + 2$

2. A triangle of area 35 square centimeters has a height which is three centimeters less than its base. Find the base and height.

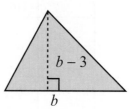

$b - 3$

b

3. The diagonal of a rectangle is ten centimeters and the width is two centimeters less than the length. Find the length and width of the rectangle.

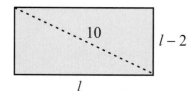

10

$l - 2$

l

4. One leg of a right triangle is four meters longer than the other. If the hypotenuse is ten meters, find the lengths of the two legs.

5. One side of a right triangle is four centimeters. The other side is seven centimeters less than twice the length of the hypotenuse. Find the length of the hypotenuse.

6. The length of a rectangle is twice the width. If each dimension is increased by three meters, the new area would be 104 square meters. Find the original dimensions.

7. Olivia is designing a picture frame. She is using wood that is one inch wide. She wants the length of the frame to be four inches more than the width. The artwork she is framing has an area of 60 square inches. Find the outer dimensions of the frame.

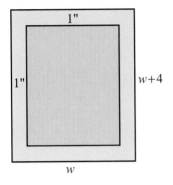

8. The Jones family has a circular pond with a diameter of eight feet. A two-foot-wide path surrounds the pond. To keep up with the Joneses, their neighbors built a larger circular pond with a four-foot-wide path. The neighbors' pond and path cover an area four times that of the Jones'. What is the diameter of the neighbors' pond?

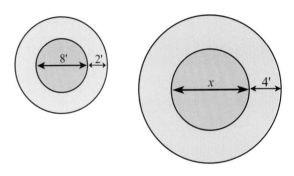

9. A polygon of n sides has $\dfrac{n(n-3)}{2}$ diagonals. How many sides does a polygon have if it has 54 diagonals?

10. An object dropped from the top of a 45-meter tower falls according to the formula $s = -5t^2 + 45$, where s represents the distance of the object above the ground at any time t in seconds. How long will it take the object to reach the ground?

11. A ball is thrown upward with a velocity of 15 meters per second. The distance s of the ball above the ground in t seconds is given by $s = 15t - 5t^2$. In how many seconds will the ball be ten meters above the ground? Why are there two answers?

12. A small motorboat can travel ten kilometers per hour in still water. The boat travels eight kilometers upstream and returns. The total time for the trip is one hour and forty minutes. What is the speed of the current?

13. An airliner traveled 1600 miles from Atlanta to Phoenix. The return trip took 48 minutes less than the original flight. If the wind speed is 50 miles per hour, find the air speed of the airliner.

14. Two pumps working together can fill a swimming pool in three hours. Working alone, pump A takes 8 hours longer than pump B to fill the pool. How many hours would it take each pump to fill the pool alone?

15. A ranch hand can mend a fence in 20 hours. If his boss works with him, they can mend the fence in one hour less time than it would take for the boss to do it alone. How long would it take to mend the fence if both worked together?

16. A farmer has 200 meters of fence on hand and wishes to enclose a rectangular field so that it will contain 2400 square meters in area. What should the dimensions of the field be?

7–6   THE DISCRIMINANT

> ### OBJECTIVES
> Upon completion of this section you should be able to:
> 1. Compute the value of the discriminant $b^2 - 4ac$.
> 2. Determine the nature of the roots of a quadratic equation from the value of the discriminant.

The solutions of equations are also called **roots**. The radicand in the quadratic formula, $b^2 - 4ac$, is called the **discriminant**. The discriminant is used to determine the nature of the roots.

If $b^2 - 4ac$ is negative, we will be finding the square root of a negative number in the quadratic formula. Therefore, there are no real solutions to the corresponding quadratic equation.

If $b^2 - 4ac = 0$, we will be finding the square root of zero in the quadratic formula. So the two solutions of the equation are equal and rational, and we can say there is one real root. Both solutions of a quadratic equation can be the same only if the quadratic is a perfect square trinomial.

If $b^2 - 4ac$ is positive, we will be finding the square root of a positive number in the quadratic formula. Therefore, the corresponding equation will have two real roots. If $b^2 - 4ac$ is a square, the roots will be rational. When the roots are rational, the original quadratic equation can be solved by factoring.

The following table summarizes the facts discussed.

Roots of a Quadratic

Value of the Discriminant	Nature of the Roots
$b^2 - 4ac < 0$	No real roots
$b^2 - 4ac = 0$	One rational root (The trinomial is a perfect square trinomial.)
$b^2 - 4ac > 0$ but not a square	Irrational and unequal roots Two different roots that contain a radical
$b^2 - 4ac > 0$ and a square	Rational and unequal roots Two different roots containing no radical (The trinomial is factorable.)

For the following examples, we will find the discriminant and determine the nature of the roots.

Example 1 $x^2 + 5x + 7 = 0$

Solution

$$b^2 - 4ac = (5)^2 - 4(1)(7)$$
$$= 25 - 28$$
$$= -3$$

So $x^2 + 5x + 7 = 0$ has no real roots. ∎

Example 2 $x^2 - 10x + 25 = 0$

Solution

$$b^2 - 4ac = (-10)^2 - 4(1)(25)$$
$$= 100 - 100$$
$$= 0$$

So $x^2 - 10x + 25 = 0$ has one rational root. ∎

Example 3 $2x^2 - 7x + 4 = 0$

Solution

$$b^2 - 4ac = (-7)^2 - 4(2)(4)$$
$$= 49 - 32$$
$$= 17$$

So $2x^2 - 7x + 4 = 0$ has two irrational roots. ∎

Example 4 $3x^2 - 8x + 4 = 0$

Solution

$$b^2 - 4ac = (-8)^2 - 4(3)(4)$$
$$= 64 - 48$$
$$= 16$$

Since the discriminant is a square, $3x^2 - 8x + 4 = 0$ has two rational roots and can be solved by factoring. ∎

EXERCISE 7-6-1

Compute $b^2 - 4ac$ for each of the following and give the nature of the roots.

1. $5x^2 - 3x + 1 = 0$ **2.** $x^2 + 4x + 4 = 0$ **3.** $2x^2 - 7x + 3 = 0$

4. $x^2 + 6x - 9 = 0$ **5.** $3x^2 + 5x + 2 = 0$ **6.** $x^2 - 4x + 4 = 0$

7. $x^2 + 5x + 7 = 0$ **8.** $2x^2 - 7x + 4 = 0$

7–7   COMPLEX NUMBERS

OBJECTIVES

Upon completion of this section you should be able to:
1. Add, subtract, multiply, and divide complex numbers.
2. Solve a quadratic equation for complex roots.

Some quadratic equations have no real solutions. However, if we expand the system of numbers to include numbers that are not classified as real numbers, then these equations have solutions.

To introduce this new set of numbers we define the **imaginary unit** i as the square root of negative 1.

DEFINITION

$$i = \sqrt{-1} \text{ or } i^2 = -1$$

Accepting this definition of i makes it possible to find values for the square roots of negative numbers, or at least to indicate such values.

Example 1   Rewrite: $\sqrt{-4}$.

Solution

$$\sqrt{-4} = \sqrt{(-1)(4)}$$
$$= \sqrt{-1}\sqrt{4}$$
$$= i\sqrt{4}$$
$$= 2i \ \blacksquare$$

To check $\sqrt{-4} = 2i$ by the definition of the square root, we evaluate $(2i)^2$.

$$(2i)^2 = (2i)(2i)$$
$$= (2)(2)(i)(i)$$
$$= 4i^2$$

But since $i^2 = -1,$

$$4i^2 = 4(-1) = -4.$$

Example 2   Rewrite: $\sqrt{-10}$.

Solution
$$\sqrt{-10} = \sqrt{(-1)(10)}$$
$$= \sqrt{-1}\sqrt{10}$$
$$= \sqrt{10}\,i \quad \blacksquare$$

A number such as $2i$ or $\sqrt{10}\,i$ is called an imaginary number.

> **DEFINITION**
>
> An **imaginary number** is a number of the form bi, where b is a real number not equal to zero and i is the imaginary unit.

EXERCISE 7-7-1

Express as imaginary numbers.

1. $\sqrt{-16}$ **2.** $\sqrt{-25}$ **3.** $\sqrt{-30}$ **4.** $\sqrt{-7}$

5. $\sqrt{-50}$ **6.** $\sqrt{-49}$ **7.** $\sqrt{-100}$ **8.** $\sqrt{-18}$

> **DEFINITION**
>
> The indicated sum of a real number and an imaginary number, $(a + bi)$ where a and b are real, is called a **complex number**.

The set of real numbers is a subset of the set of complex numbers. A real number a can be expressed as $a + 0i$ and therefore, by definition, is complex. Also, the set of imaginary numbers is a subset of the set of complex numbers since an imaginary number bi can be expressed as $0 + bi$.

Complex numbers can be added, subtracted, multiplied, raised to powers, and so on, as you would expect of any set of numbers.

RULE

To add or subtract complex numbers, combine the real parts of each number and combine the imaginary parts of each number.

Note that this rule follows a previous rule that only like terms can be combined.

Example 3   Add: $(7+6i)+(3+2i)$

Solution

$$(7+3)+(6i+2i)$$
$$10+8i \ \blacksquare$$

Example 4   Add: $(3+4i)-(2-6i)$

Solution

$$(3-2)+(4i+6i)$$
$$1+10i \ \blacksquare$$

RULE

To multiply complex numbers, use the distributive property.

Example 5   Multiply: $5(3+4i)$

Solution

$$5(3)+5(4i)$$
$$15+20i \ \blacksquare$$

When both complex numbers are in the form $a + bi$, we use the FOIL method.

Example 6   Multiply: $(3+2i)(4+6i)$

Solution

$$12+18i+8i+12i^2$$
$$12+26i+12(-1)$$
$$26i \ \blacksquare$$

Example 7   Expand: $(3+i)^2$

Solution

$$(3+i)(3+i)$$
$$9+3i+3i+i^2$$
$$9+6i+(-1)$$
$$8+6i \ \blacksquare$$

Division is defined as multiplying by the reciprocal of the divisor.

Example 8   Divide: $(3+4i)\div(2+i)$

Solution
$$(3+4i)\cdot\frac{1}{(2+i)}=\frac{3+4i}{2+i}$$

If we write this as $\dfrac{3+4\sqrt{-1}}{2+\sqrt{-1}}$, you will recognize it as a problem in which we need to rationalize the denominator. We should multiply the numerator and denominator by $2-i$ (known as the *conjugate* of $2+i$).

$$\frac{3+4i}{2+i}\cdot\frac{2-i}{2-i}=\frac{6+5i-4i^2}{4-i^2}$$

$$=\frac{10+5i}{5}$$

This can be further simplified.
$$=\frac{5(2+i)}{5}$$

$$=2+i \quad\blacksquare$$

The **conjugate** of the complex number $(a+bi)$ is $(a-bi)$.

RULE

To divide complex numbers, multiply by the reciprocal of the divisor. If i occurs in the denominator, multiply both the numerator and denominator by the conjugate of the denominator.

Example 9   Divide: $i\div(1-i)$

Solution
$$i\cdot\frac{1}{1-i}=\frac{i}{1-i}\cdot\frac{1+i}{1+i}$$

$$=\frac{i+i^2}{1-i^2}$$

$$=\frac{i-1}{2}$$

$$=-\frac{1}{2}+\frac{1}{2}i \quad\blacksquare$$

EXERCISE 7-7-2

Perform the operations.

1. $(2+3i)+(5+4i)$ 2. $(6+5i)+(1-6i)$ 3. $(5-4i)+2(4+2i)$

4. $(5-2i)-(8-i)$ 5. $(11+7i)-(3+4i)$ 6. $(11+i)-3(2-5i)$

7. $(2+3i)(1+4i)$ 8. $(x+iy)(x-iy)$ 9. $(5+4i)^2$

10. $(6-5i)^2$ 11. $(2+3i)\div(1-2i)$ 12. $5\div(4+3i)$

13. $2i\div(6+i)$ 14. $(3+i)\div i$ 15. $(1-4i)\div(2-3i)$

The most common use of complex numbers at this level of algebra is in expressing solutions to quadratic equations that have no real solution.

Example 10   Solve and check: $x^2-2x+5=0$

Solution   Since the expression on the left will not factor, we will use the quadratic formula with $a=1$, $b=-2$, and $c=5$.

$$x = \frac{2\pm\sqrt{(-2)^2-4(1)(5)}}{2}$$

$$= \frac{2\pm\sqrt{-16}}{2}$$

$$= \frac{2\pm 4i}{2}$$

$$= \frac{2(1\pm 2i)}{2}$$

$$= 1\pm 2i$$

Check $x = 1 + 2i$ *Check* $x = 1 - 2i$

$$(1+2i)^2 - 2(1+2i) + 5 = 0 \qquad\qquad (1-2i)^2 - 2(1-2i) + 5 = 0$$

$$1 + 4i + 4i^2 - 2 - 4i + 5 = 0 \qquad\qquad 1 - 4i + 4i^2 - 2 + 4i + 5 = 0$$

$$4 + 4i^2 = 0 \qquad\qquad\qquad\qquad 4 + 4i^2 = 0$$

$$4 + 4(-1) = 0 \qquad\qquad\qquad\qquad 4 + 4(-1) = 0$$

$$0 = 0 \qquad\qquad\qquad\qquad\qquad 0 = 0$$

We see that both solutions check. Thus the solution set is $\{1 \pm 2i\}$. ∎

EXERCISE 7-7-3

Solve.

1. $x^2 - 4x + 8 = 0$ **2.** $2x^2 - 2x + 1 = 0$ **3.** $8x^2 - 4x + 1 = 0$

4. $x^2 + x + 1 = 0$ **5.** $x^2 - x + 1 = 0$ **6.** $x^2 - 2x + 4 = 0$

7. $3x^2 + x + 1 = 0$ **8.** $x^2 - 4x + 5 = 0$ **9.** $5x^2 - 2x + 3 = 0$

7–8   EQUATIONS WITH RADICALS

OBJECTIVES

Upon completion of this section you should be able to:
1. Solve equations involving radicals.
2. Identify extraneous roots.

Consider the two equations $x = 2$ and $x^2 = 4$. These equations certainly have something in common but it is also obvious that they are not equivalent, since 2 is the only solution to $x = 2$ but 2 and -2 are both solutions to $x^2 = 4$. If both sides of an equation are squared, the resulting equation is not always equivalent to the original equation. However, any solution to the original equation will be a solution of the resulting equation.

In general, if both sides of an equation are raised to the same power, the resulting equation will contain all solutions of the original equation. However, the resulting equation may also contain solutions that are *not* solutions of the original equation. These solutions are called **extraneous roots**. It is therefore *necessary* to check all solutions in the original equation.

A PROCEDURE FOR SOLVING EQUATIONS INVOLVING SQUARE ROOTS
1. Isolate the radical if possible.
2. Square both sides of the equation.
3. If necessary, repeat steps 1 and 2 until all radicals have been eliminated.
4. Solve the resulting equation.
5. Check for extraneous roots.

Example 1   Solve: $x - \sqrt{2x-5} = 4$

Solution   First we isolate the radical on one side of the equation.

$$-\sqrt{2x-5} = 4 - x$$

Then we square each side of the equation.

$$\left(-\sqrt{2x-5}\right)^2 = (4-x)^2$$
$$2x-5 = 16 - 8x + x^2$$
$$0 = x^2 - 10x + 21$$
$$0 = (x-7)(x-3)$$
$$x-7=0 \quad \text{or} \quad x-3=0$$
$$x=7 \qquad\qquad x=3$$

Check $\quad\quad\quad x = 7 \quad\quad\quad\quad\quad\quad\quad\quad x = 3$

$$7 - \sqrt{2(7) - 5} = 4 \quad\quad\quad\quad 3 - \sqrt{2(3) - 5} = 4$$

$$7 - \sqrt{9} = 4 \quad\quad\quad\quad\quad\quad 3 - \sqrt{1} = 4$$

$$7 - 3 = 4 \quad\quad\quad\quad\quad\quad\quad 3 - 1 = 4$$

$$4 = 4 \quad\quad\quad\quad\quad\quad\quad\quad 2 = 4$$

Therefore, 7 is a solution and 3 is *not* a solution. The solution set is $\{7\}$. ∎

Example 2   Solve: $\sqrt{x+1} - \sqrt{3x} = -1$

Solution   Here we have more than one radical in the equation. We will isolate one of the radicals by adding $\sqrt{3x}$ to both sides. Then when we square both sides, we will eliminate the radical $\sqrt{x+1}$.

$$\sqrt{x+1} = \sqrt{3x} - 1$$

$$\left(\sqrt{x+1}\right)^2 = \left(\sqrt{3x} - 1\right)^2$$

$$x + 1 = \left(\sqrt{3x} - 1\right)\left(\sqrt{3x} - 1\right)$$

$$x + 1 = \left(\sqrt{3x}\right)^2 - 2\sqrt{3x} + 1$$

$$x + 1 = 3x - 2\sqrt{3x} + 1$$

$$2\sqrt{3x} = 2x$$

$$\sqrt{3x} = x$$

$$\left(\sqrt{3x}\right)^2 = x^2$$

$$3x = x^2$$

$$0 = x^2 - 3x$$

$$0 = x(x - 3)$$

$$x = 0 \quad \text{or} \quad x - 3 = 0$$

$$x = 3$$

Check $\quad\quad\quad x = 0 \quad\quad\quad\quad\quad\quad\quad\quad x = 3$

$$\sqrt{0+1} - \sqrt{3(0)} = -1 \quad\quad\quad \sqrt{3+1} - \sqrt{3(3)} = -1$$

$$1 - 0 = -1 \quad\quad\quad\quad\quad\quad 2 - 3 = -1$$

$$1 = -1 \quad\quad\quad\quad\quad\quad\quad -1 = -1$$

Therefore, 0 is not a solution and 3 is a solution. The solution set is {3}. ∎

EXERCISE 7-8-1

Solve.

1. $\sqrt{x} = 6$

2. $\sqrt{x-1} = 7$

3. $\sqrt{3a+2} = 5$

4. $\sqrt{x} = 6 - x$

5. $2\sqrt{x} = x - 3$

6. $\sqrt{2y-5} = 10 - y$

7. $p - 2\sqrt{2p+1} = -2$

8. $5 - \sqrt{5x-1} = x$

9. $\sqrt{5x+1} = 1 + \sqrt{3x}$

10. $\sqrt{x+7} = 1 + \sqrt{2x}$

11. $\sqrt{x-5} - \sqrt{x+3} = 2$

12. $\sqrt{3y+10} - \sqrt{2y-1} = -2$

7–9   EQUATIONS QUADRATIC IN FORM

OBJECTIVES

Upon completion of this section you should be able to:
1. Determine if an equation is quadratic in form.
2. Solve equations that are quadratic in form.

An equation is **quadratic in form** if a suitable substitution for the variable can be found so that the resulting equation is a quadratic equation. For example,

$$x^4 - 5x^2 + 6 = 0$$

is quadratic in form since the substitution $u = x^2$ would give

$$u^2 - 5u + 6 = 0,$$

which is a quadratic equation in the variable u. The original equation may be solved by solving the resulting quadratic equation and then using these solutions and the substitution to obtain the solution to the original equation.

Example 1   Solve: $x^4 - 5x^2 + 6 = 0$.

Solution   Let $u = x^2$. Then $u^2 = x^4$.

$$u^2 - 5u + 6 = 0$$
$$(u - 3)(u - 2) = 0$$

$$u - 3 = 0 \quad \text{or} \quad u - 2 = 0$$
$$u = 3 \qquad\qquad u = 2$$

Since $u = x^2$, substitute x^2 for u and solve.

$$x^2 = 3 \quad \text{or} \quad x^2 = 2$$
$$x = \pm\sqrt{3} \qquad x = \pm\sqrt{2}$$

Checking these values in the original equation, we find they all satisfy the equation. Therefore, the solution set is $\left\{ -\sqrt{3}, -\sqrt{2}, \sqrt{2}, \sqrt{3} \right\}$. ■

Note that this fourth-degree equation has four roots. An equation of degree n will have n complex roots.

Example 2   Solve: $x - 4x^{\frac{1}{2}} + 3 = 0$

Solution   Let $u = x^{\frac{1}{2}}$. Then $u^2 = \left(x^{\frac{1}{2}}\right)^2 = x$.

$$u^2 - 4u + 3 = 0$$

$$(u - 3)(u - 1) = 0$$

$$u = 3 \quad \text{or} \quad u = 1$$

$$x^{\frac{1}{2}} = 3 \quad \text{or} \quad x^{\frac{1}{2}} = 1$$

$$\sqrt{x} = 3 \quad \text{or} \quad \sqrt{x} = 1$$

$$x = 9 \quad \text{or} \quad x = 1$$

By substituting each of these values into the original equation, we find that 9 is a solution and 1 is not a solution. So the solution set is {9}. ∎

Example 3   Solve: $x^{-2} + 6x^{-1} + 8 = 0$

Solution   Let $u = x^{-1}$. Then $u^2 = x^{-2}$.

$$u^2 + 6u + 8 = 0$$

$$(u + 2)(u + 4) = 0$$

$$u = -2 \quad \text{or} \quad u = -4$$

$$x^{-1} = -2 \quad \text{or} \quad x^{-1} = -4$$

$$\frac{1}{x} = -2 \qquad \frac{1}{x} = -4$$

$$1 = -2x \qquad 1 = -4x$$

$$-\frac{1}{2} = x \qquad -\frac{1}{4} = x$$

By substituting each of these values into the original equation we find that both are solutions. So the solution set is $\left\{-\dfrac{1}{2}, -\dfrac{1}{4}\right\}$. ∎

EXERCISE 7–9–1

Solve.

1. $x^4 - 5x^2 + 4 = 0$

2. $x^4 - 10x^2 + 9 = 0$

3. $x^4 - 6x^2 + 8 = 0$

4. $x^{-2} + x^{-1} - 6 = 0$

5. $2x^{-2} - 5x^{-1} + 3 = 0$

6. $8z^{-2} + 10z^{-1} + 3 = 0$

7. $x - 6x^{\frac{1}{2}} + 8 = 0$

8. $x - 4x^{\frac{1}{2}} - 21 = 0$

9. $a - 2a^{\frac{1}{2}} = 15$

CHAPTER 7 SUMMARY

The number in brackets refers to the section of the chapter that discusses the concept.

Terminology

- A **quadratic equation** is a second-degree equation in one variable. [7–1]
- The **standard form** of a quadratic equation is $ax^2 + bx + c = 0$, $a \neq 0$. [7–1]
- An **incomplete quadratic equation** is an equation of the form $ax^2 + bx = 0$ or $ax^2 + c = 0$, $a \neq 0$. [7–2]
- **Completing the square** is a process for solving quadratic equations. [7–3]
- The **quadratic formula** is a formula used to solve quadratic equations.
$$x = \frac{-b \pm \sqrt{b^2 - 4ac}}{2a} \quad [7-4]$$
- In the quadratic formula, the radicand $b^2 - 4ac$ is called the **discriminant**. [7–6]
- A **complex number** is a number of the form $a + bi$, where a and b are real numbers and $i = \sqrt{-1}$. [7–7]
- The **conjugate** of the complex number $a + bi$ is $a - bi$. [7–7]
- An **extraneous root** is a root derived in the solution process that does not solve the original equation. Extraneous roots may occur when both sides of an equation are raised to a power. [7–8]
- An equation is **quadratic in form** if a suitable substitution for the variable can be found so that the resulting equation is quadratic. [7–9]

Rules and Procedures

Solving Quadratic Equations

- To solve a quadratic equation by factoring, write the equation in standard form, factor, and then set each factor equal to zero. [7–1]

- To solve a quadratic equation of the form $ax^2 + c = 0$, solve for x^2 and then find the square root of both sides. [7–2]
- To solve a quadratic equation by completing the square, complete the following steps. [7–3]

  1. Write the equation in standard form.
  2. If the coefficient of x^2 is not 1, divide each side of the equation by that coefficient.
  3. Add and subtract terms so that the x^2 and x terms are on one side of the equation and the numerical term is on the other side.
  4. Complete the square and add this number to both sides of the equation.
  5. Factor the completed square.
  6. Take the square root of both sides of the equation, remembering that if $A^2 = B$, then $A = \pm\sqrt{B}$.
  7. Solve for the two values of x.

- To solve a quadratic equation using the quadratic formula, first write the equation in standard form $ax^2 + bx + c = 0$. Then substitute the values of a, b, and c into the formula.

$$x = \frac{-b \pm \sqrt{b^2 - 4ac}}{2a} \quad [7-4]$$

Roots of a Quadratic Equation

- The nature of the roots of a quadratic equation is dependent on the value of the discriminant. If the value of the discriminant is greater than or equal to zero, the roots are real. If the value is negative, there are no real roots. [7–6]

Complex Numbers

- To add or subtract complex numbers, combine the real parts of each number and combine the imaginary parts of each number. [7–7]
- To multiply complex numbers, use the distributive property. [7–7]
- To divide complex numbers, multiply by the reciprocal of the divisor. If i occurs in the denominator, multiply both the numerator and denominator by the conjugate of the denominator. [7–7]

Radical Equations

- To solve equations containing square roots, complete the following steps. [7–8]

1. Isolate the radical if possible.
2. Square both sides of the equation.
3. If necessary, repeat step 1 and step 2 until all radicals have been eliminated.
4. Solve the resulting equation.
5. Check for extraneous roots.

Equations Quadratic in Form

- Equations that are quadratic in form are solved by finding a suitable substitution to create a quadratic equation, solving the quadratic equation and then back-substituting to find the value(s) of the original variable. [7–9]

CHAPTER 7 REVIEW

Solve by factoring.

1. $x^2 + x = 30$

2. $r^2 - 11r + 24 = 0$

3. $6x^2 + 7x = 3$

Solve.

4. $s^2 = 17$

5. $5x^2 = 100$

6. $9z^2 = 6z$

Solve by completing the square.

7. $x^2 + 8x + 15 = 0$ **8.** $p^2 - 10p + 8 = 0$ **9.** $x^2 + 3x - 1 = 0$

Solve by using the quadratic formula.

10. $x^2 + 3x - 28 = 0$ **11.** $2x^2 + 8x + 7 = 0$ **12.** $x^2 - x + 1 = 0$

Compute $b^2 - 4ac$ for each of the following and give the nature of the roots.

13. $x^2 + 3x = 2$ **14.** $x^2 - x - 1 = 0$ **15.** $2m^2 - 3m + 2 = 0$

Perform the indicated operation.

16. $(3 - 2i) + (1 + 5i)$ **17.** $(6 + i) - (3 - 2i)$ **18.** $(2 + 5i)(1 - 3i)$

19. $(7 + i) \div (2 + i)$ **20.** $3i \div (4 + 2i)$

Solve for real or complex roots.

21. $x(6x-5)=4$

22. $x^2+5=4x$

23. $\dfrac{1}{2}p^2+p=\dfrac{1}{4}$

24. $2-\dfrac{5}{x^2}=\dfrac{3}{x}$

25. $x^4-10x^2+9=0$

26. $y+3y^{\frac{1}{2}}=4$

27. $1-9x^{-1}+20x^{-2}=0$

28. $a-3\sqrt{a}=-2$

29. $4+\sqrt{2x}=x$

30. $\sqrt{y+1}-\sqrt{y-3}=2$

31. The area of a rectangle is 105 square meters. Find the dimensions if the length is one meter more than twice the width.

32. One leg of a right triangle is four meters more than twice the other leg. If the hypotenuse is 26 meters, find the lengths of the legs.

33. An airliner has an air speed of 400 miles per hour. It travels 1920 miles with the wind from San Francisco to Indianapolis. After refueling, it returns to San Francisco. If the total time is flight is ten hours, what is the wind speed?

34. If Bill and Bob work together they can landscape a garden in 4 hours. Bill can landscape the garden by himself in 6 hours less time than it takes Bob to do it alone. How many hours would it take Bill to do the job alone?

CHAPTER 7 PRACTICE TEST

1. Compute $b^2 - 4ac$ for $2x^2 - 3x - 1 = 0$ and give the nature of the roots.

1. _____

2. Find the value of c that will make the following trinomial a perfect square trinomial: $x^2 - 3x + c$

2. _____

Solve for real or complex roots:

3. $x^2 = 10$

3. _____

4. $3x^2 - 12 = 0$

4. _____

5. $y^2 - 7y = 0$

5. _____

6. $3t^2 + 6t - 9 = 0$

6. _____

7. $x(x-4) = -1$

7. _____

8. $2z^2 + 5z = 3$

8. _____

9. $x^2 - 3x + 4 = 0$

9. _____

10. $x^4 - 3x^2 - 10 = 0$

10. _____

11. $x - \sqrt{2x} = 12$

11. _____

12. $\sqrt{x+5} + \sqrt{x} = 5$

12. _____

13. The area of a rectangle is 84 square meters. If the length is 5 meters greater than the width, find the width.

13. _____

CHAPTER 8

SURVEY

The following questions refer to material discussed in this chapter. Work as many problems as you can and check your answers with the answer key in the back of the book. The results will direct you to the sections of the chapter in which you need to work. If you answer all questions correctly, you may already have a good understanding of the material contained in this chapter.

1. Sketch the graph of $y = x^2 + 2x - 3$.

1.

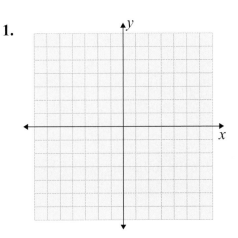

2. Which of the following equations gives y as a function of x?
Give the domain of the function.
   a. $y = \sqrt{x+1}$ **b.** $y^2 = x - 5$

2. _____

3. Find the slope of the line through the points $(2, -1)$ and $(-2, 3)$.

3. _____

4. Find the standard form of the equation of the line through the points $(3, -1)$ and $(-4, 2)$.

4. _____

5. Find the standard form of the equation of the line with slope $m = 2$ and passing through the point $(-3, 1)$.

5. _____

6. Find the standard form of the equation of the line that is perpendicular to $2x + 3y = 1$ and passes through the point $(-3, 1)$.

6. _____

7. Solve the following system by graphing:
$$\begin{cases} y = x^2 - 2x + 3 \\ x + y = 5 \end{cases}$$

7.

8. Solve algebraically: $\begin{cases} 2x + 3y = 6 \\ x - 2y = -11 \end{cases}$

8. _____

9. Sketch the graph of the following linear inequality: $2x + y < 5$

9.

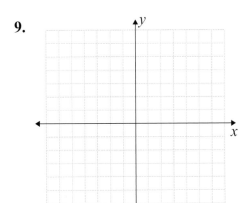

10. Solve the following system of inequalities by graphing: $\begin{cases} 3x - 2y \le 6 \\ x + 2y \ge 4 \end{cases}$

10.

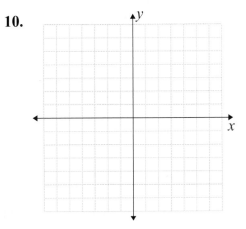

11. Solve: $\begin{cases} 3x + 4y + z = -2 \\ y + z = 1 \\ 2x - y - z = -5 \end{cases}$

11. _____

CHAPTER 8

Relations and the Cartesian Coordinate System

In Chapters 2 and 7 we discussed equations containing one variable. In this chapter we will discuss equations containing more than one variable as well as techniques for solving systems of these equations.

8–1   CARTESIAN COORDINATE SYSTEM AND GRAPHING EQUATIONS IN TWO VARIABLES

> **OBJECTIVES**
>
> Upon completion of this section you should be able to:
> 1. Locate points in the Cartesian coordinate system.
> 2. Sketch the graph of a relation by plotting points.

If we are given two sets of numbers, it may be possible to establish a rule that shows a relationship between elements of the first set and elements of the second set.

Suppose we start with two sets of numbers

A: {1, 3, 5}   and
B: {2, 4, 6, 8, 10}.

There are many ways to relate some of the numbers in set A to numbers in set B. Each result is a pair of numbers, one from set A and one from set B. Examples 1 and 2 give two possibilities.

Example 1   Find the pairs of numbers that fit the rule: "Relate a number in set A to a number in set B so that the number in set B is twice as large as the number in set A."

Solution

1 is related to 2
3 is related to 6
5 is related to 10 ■

The results in Example 1 can be written as a set of ordered pairs, {(1, 2), (3, 6), (5, 10)}, where the first number in each pair is from set A and the second number is from set B. A pair of numbers where the order of the numbers matters is known as an **ordered pair**. One reason the ordered pair (2, 1) would not be part of the solution to Example 1 is that 2 is not in set A.

Example 2   Given the same two sets A and B, find the ordered pairs of numbers that fit the rule: "Relate a number in set A to a number in set B so that the number from set B is larger than the number from set A."

Solution   1 is related to 2, 4, 6, 8, and 10
3 is related to 4, 6, 8, and 10
5 is related to 6, 8, and 10

Writing this solution as a set of ordered pairs, we have

$$\{(1,2),(1,4),(1,6),(1,8),(1,10),(3,4),(3,6),(3,8),(3,10),(5,6),(5,8),(5,10)\}. \;■$$

DEFINITION
A **relation** is a set of ordered pairs.

The numbers in an ordered pair are called the **coordinates** of the ordered pair. The first number in an ordered pair is known as the first coordinate or **abscissa**. The second number in an ordered pair is known as the second coordinate or **ordinate**.

In mathematics, the rule relating the elements of the first set of numbers to elements of the second set of numbers is usually given as an algebraic equation. For instance, in Example 1 if x represents an element of set A, then $2x$ represents the related element in set B. This rule could be written as the equation $y = 2x$, where y represents the corresponding values in set B. Notice that this equation contains two variables, x and y. The variable to which values are assigned is called the **independent variable** and the other is called the **dependent variable**. For $y = 2x$, the independent variable is x and the dependent variable is y.

Example 3   Given the same two sets A and B, find the set of ordered pairs that belong to the relation "Relate a number in set A to a number in set B so that the number from set B is one larger than the number in set A." Write an algebraic equation for this relation with x as the independent variable and y as the dependent variable.

Solution $\{(1, 2), (3, 4), (5, 6)\}$
$y = x + 1$ ■

EXERCISE 8-1-1

A: {1, 3, 5}, B: {2, 4, 6, 8, 10}
Find the set of ordered pairs for each relation and write an algebraic equation for the relation.

1. Relate a number in set A to a number in set B so that the number from set B is three more than the number in set A.

2. Relate a number in set A to a number in set B so that the number from set B is one less than the number in set A.

Now let set A contain all real numbers and set B also contain all real numbers. Suppose we want to find all the ordered pairs of numbers that belong to the relation "Relate a number in set A to a number in set B so that the number in set B is twice as large as the number in set A" or, algebraically, $y = 2x$. We could begin by choosing some values from set A for x and pairing them with the corresponding values from set B. However, it quickly becomes clear that since we will use all real numbers, we would never finish the process. We cannot list the solution as a set of ordered pairs. We must look for an alternative way to show the solution to the relation. This alternative is to graph the solution.

One method of uniquely naming each point in the plane is the **rectangular** or **Cartesian coordinate system** (named for the French mathematician René Descartes, 1596-1650). This rectangular coordinate system is constructed with two mutually perpendicular real number lines in the plane that intersect at zero on each line. One line is vertical with positive direction upward and the other is horizontal with positive direction to the right. The two lines are called **coordinate axes** and their point of intersection is called the **origin**. The horizontal axis is usually referred to as the **x-axis** and the vertical axis is usually referred to as the **y-axis**.

The four regions of the plane formed by the x-axis and y-axis are called **quadrants** and are numbered in a counterclockwise direction starting with the upper right.

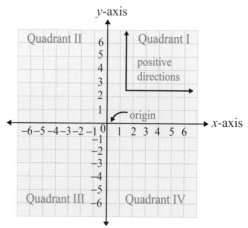

Any point in the plane can be located by moving first along the *x*-axis and then parallel to the *y*-axis. We use an ordered pair of numbers such as (x, y) to represent a point in the plane. The first number (abscissa) represents the distance and direction (right or left) from the origin along the *x*-axis and the second number (ordinate) represents the distance and direction (up or down) parallel to the *y*-axis. An ordered pair of numbers represents one and only one point on the plane. In the Cartesian coordinate system, points on the plane are always represented by an ordered pair (x, y).

EXERCISE 8-1-2

1. Locate each of the following points on the coordinate plane. Label each point with the letter and the ordered pair. *A* is given as an example. *A*: $(-3, 5)$ *B*: $(5, 2)$ *C*: $(5, -4)$ *D*: $(-4, -5)$ *E*: $(0, 4)$ *F*: $(5, 0)$ *G*: $(-3, 0)$ *H*: $(0, -3)$

2. Note the points indicated on the coordinate plane. Give the ordered pair associated with each point.

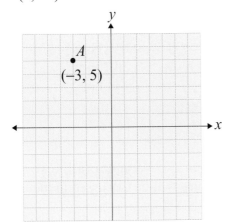

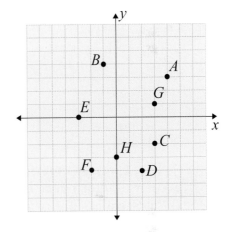

The graph of a relation between *x* and *y* can be represented in the coordinate system. We do this by expressing the relation as ordered pairs and locating points on the plane represented by these ordered pairs. It is generally not possible to locate all points given by an algebraic equation. However, each algebraic equation in two variables forms a unique pattern of points. To sketch the graph, we locate a sufficient number of points to establish the pattern, and then complete the graph by filling in the missing points.

Example 4   Sketch the graph of $y = 3x - 1$.

Solution   First we must find ordered pairs (x, y) that are solutions to the given equation. We accomplish this by substituting values of x in the equation and solving for corresponding values of y.

If $x = -1$, then $y = 3(-1) - 1 = -4$. Therefore, the ordered pair $(-1, -4)$ is a solution.
If $x = 0$, then $y = 3(0) - 1 = -1$. Therefore, $(0, -1)$ is a solution.
If $x = 1$, then $y = 3(1) - 1 = 2$. Therefore, $(1, 2)$ is a solution.
If $x = 2$, then $y = 3(2) - 1 = 5$. Therefore, $(2, 5)$ is a solution.

We now locate the points $(-1, -4)$, $(0, -1)$, $(1, 2)$, and $(2, 5)$ on the coordinate plane. These points establish a pattern that is a line. We fill in the rest of the ordered pairs to complete the solution by drawing a line. We say that the solution to $y = 3x - 1$ is a line.

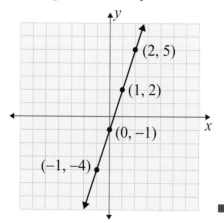

The graph of any first-degree equation in *two* variables (a **linear equation**) will form a line. Linear equations in two variables will be discussed in more detail in Section 8–3.

CAUTION   When sketching the graph of an equation in two variables, it is necessary to find enough ordered pairs to establish a pattern. If too few points are used, the sketch will be inaccurate. When sketching the graph of a line (a first-degree equation), only two points are necessary since two points determine a line. However, it is best to find at least three points, using any extra points as a check.

One convenient way to keep track of the solutions to the equation is to place the ordered pairs in a table called a **table of values**. A table of values can be written horizontally or vertically. Either of the following could be used for Example 4.

x	−1	0	1	2
y	−4	−1	2	5

x	y
−1	−4
0	−1
1	2
2	5

Example 5   Sketch the graph of $y = -2x + 1$.

Solution   First choose values for x. Then find the corresponding values of y to form a table of values.

x	y
−1	3
0	1
1	−1
2	−3

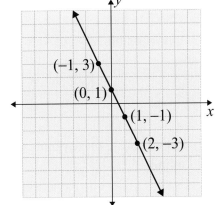

DEFINITION

Points where a graph intersects the x-axis are called **x-intercepts** and points where a graph intersects the y-axis are called **y-intercepts**. An x-intercept will be of the form $(x, 0)$ and a y-intercept will be of the form $(0, y)$.

The x- and y-intercepts are usually not too difficult to find and are generally indicated on the graph of a relation. Every point on the x-axis has an ordinate of 0. Every point on the y-axis has an abscissa of 0. So we use the following rule to find the intercepts of an equation.

RULE

1. To find an x-intercept, let $y = 0$ and solve for x.
2. To find a y-intercept, let $x = 0$ and solve for y.

In Example 5, the y-intercept is $(0, 1)$ and the x-intercept is $\left(\dfrac{1}{2}, 0\right)$.

Example 6   Sketch the graph of $y = x^2 - 3x - 4$.

Solution   We again assign values to x and find the corresponding values of y.

If $x = -2$, then $y = (-2)^2 - 3(-2) - 4 = 6$.

Therefore, $(-2, 6)$ represents a point on our graph.

If $x = 0$, then $y = (0)^2 - 3(0) - 4 = -4$.

Therefore, $(0, -4)$ is the y-intercept on our graph.

To find the x-intercepts, let $y = 0$ and solve $0 = x^2 - 3x - 4$. This is a quadratic equation that can be solved by factoring. The x-intercepts are $(-1, 0)$ and $(4, 0)$.

We continue this process until a sufficient number of points have been found to establish a pattern. The number of points necessary will vary with the relation. However, we should choose a wide variety of values of x to help ensure that we obtain more than just a "local" pattern.

One possible table of values for $y = x^2 - 3x - 4$ is:

x	-2	-1	0	1	2	3	4	5
y	6	0	-4	-6	-6	-4	0	6

As we locate these points on the coordinate plane, it becomes clear that they are not on a line.

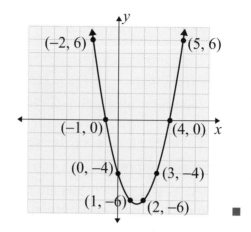

The curve in the preceding graph is a **parabola**. The curves formed by second-degree polynomial equations in two variables are the **circle, ellipse, parabola,** and **hyperbola**. These are studied in detail in a course in analytic geometry.

Sometimes the relation might be such that we wish to choose values of y and solve for values of x.

Example 7   Sketch: $x = y^2 + 2y - 1$

Solution   First find a table of values by assigning values to y.
If $y = -1$, then $x = (-1)^2 + 2(-1) - 1 = -2$.
Therefore, $(-2, -1)$ is a solution.
If $y = 0$, then $x = (0)^2 + 2(0) - 1 = -1$.
Therefore, $(-1, 0)$ is a solution.

Continuing to choose values for y, we might obtain the following table.

x	7	2	-1	-2	-1	2	7
y	-4	-3	-2	-1	0	1	2

Plotting these points gives us the following graph.

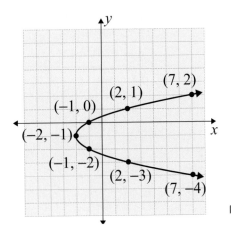

EXERCISE 8-1-3

Sketch the graphs.

1. $y = x$

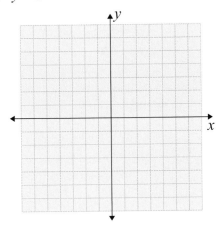

2. $y = x + 3$

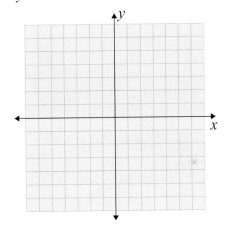

3. $y = x - 4$

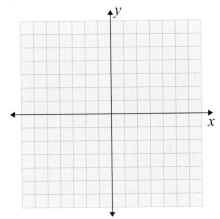

4. $y = 4 - x$

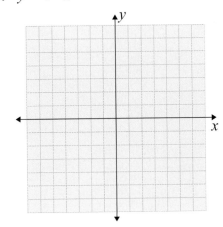

5. $y = 2x - 3$

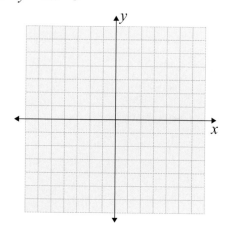

6. $y = 3 - 2x$

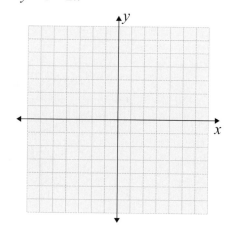

7. $x + 3y = 2$

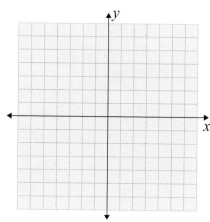

8. $3x - 2y = 7$

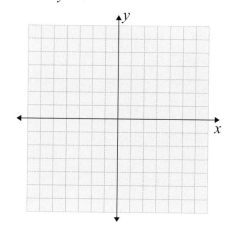

9. $y = x^2$

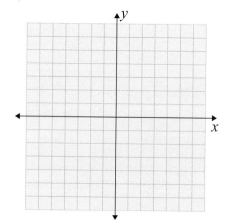

10. $y = x^2 + 2x - 3$

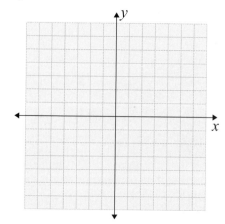

11. $y = x^2 - 4x + 3$

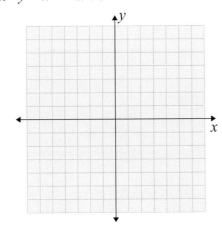

12. $y = -x^2 + 2x - 1$

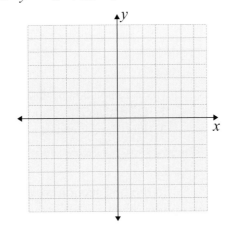

13. $x = y^2 - 4y + 2$

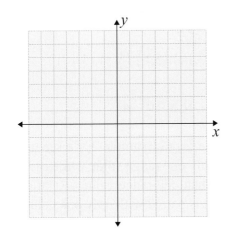

14. $x = -2y^2 + 8y - 5$

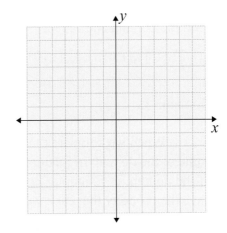

15. $y = \sqrt{x}$

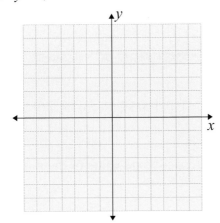

16. $y = x^3$

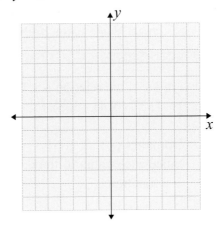

8–2   FUNCTIONS

> **OBJECTIVES**
>
> Upon completion of this section you should be able to:
> 1. Determine if a relation is a function.
> 2. Find the domain of a function.

In the first section we defined a relation as a set of ordered pairs. The set of all first numbers (abscissas) in the ordered pairs of a relation is called the **domain**. The set of second numbers (ordinates) in the ordered pairs of a relation is called the **range**.

Example 1   Find the domain and range: $\{(1, 2), (3, 6), (5, 10)\}$

Solution   Domain: $\{1, 3, 5\}$ Range: $\{2, 6, 10\}$ ■

Example 2   Find the domain and range:

$$\{(1,2),(1,4),(1,6),(1,8),(1,10),(3,4),(3,6),(3,8),(3,10),(5,6),(5,8),(5,10)\}$$

Solution   Domain: $\{1, 3, 5\}$ Range: $\{2, 4, 6, 8, 10\}$ ■

When we compare the domains in Example 1 and Example 2, we see that each value of the domain in Example 1 is paired with only one range value while each value of the domain in Example 2 is paired with more than one range value. Relations that have the property that each value from the domain is paired with only one value of the range are classified as *functions*. Therefore, the relation in Example 1 is classified as a function and the relation in Example 2 is not a function.

> **DEFINITION**
>
> A **function** is a relation such that each value of the domain is paired with exactly one value of the range.

Example 3   Determine if the relation is a function.
   a. $\{(-1, 3), (0, 4), (3, 3), (5, -7)\}$ **b.** $\{(-1, 3), (-1, -3), (2, 5), (3, 7)\}$

Solution   **a.** Domain: $\{-1, 0, 3, 5\}$ Range: $\{-7, 3, 4\}$
   Since each value of the domain is paired with exactly one value of the range, this is a function.

   b. Domain: $\{-1, 2, 3\}$ Range: $\{-3, 3, 5, 7\}$
   Since the value -1 from the domain is paired with two values of the range, -3 and 3, this is not a function. ■

Recall that a relation can be given as an algebraic equation in two variables. One of the variables is the independent variable and its values would be from the domain. The other variable is the dependent variable and its values would be from the range. Therefore, an algebraic equation can be classified as a function if each value of the independent variable is paired with exactly one value of the dependent variable. If the independent variable is x and the dependent variable is y, we say that y is a function of x.

Example 4   Determine if y is a function of x and find the domain: $y = x^2 + 2x - 5$

Solution   From the definition, y is a function of x if for each value of x there is exactly one value of y. We can determine this by substituting values for x (the independent variable) and finding the corresponding values of y (the dependent variable).

If $x = 3$, then $y = (3)^2 + 2(3) - 5 = 10$.

We note that if any number is squared and then added to twice that number, and then 5 is subtracted, we will arrive at a unique value. We conclude that y is a function of x because each x yields exactly one y. To find the domain, we must find the largest set of numbers that can be used for x to give valid values for y. One way to determine this is to ask, "Are there any numbers that cannot be substituted for x?" In this case, we know that any real number can be squared, the result can be added to twice that number, and that result can be reduced by 5. So we conclude that the domain of $y = x^2 + 2x - 5$ is the set of all real numbers. ∎

Example 5   Determine if y is a function of x and find the domain: $y = \sqrt{x}$

Solution   Since the radical represents only the principal square root, a value substituted for x will yield only one value of y. Therefore, we conclude that y is a function of x.
To determine the domain we must ask, "Does $\sqrt{x}$ yield a real number for all values of x?" We know that $\sqrt{x}$ is not a real number when $x < 0$. Thus the domain is zero and all positive numbers. We can write this as $x \geq 0$. ∎

CAUTION   Be alert for restrictions in the domain when a function involves radicals or fractions.

Example 6   Determine if y is a function of x and find the domain: $y = \dfrac{1}{x-2}$

Solution   If we substitute a number for x, can $\dfrac{1}{x-2}$ have more than one value? No. Therefore, y is a function of x. To determine the domain, we know that zero cannot occur as the denominator of a fraction. If $x = 2$, then $\dfrac{1}{x-2}$ becomes $\dfrac{1}{0}$. Thus the domain of this function is the set of all real numbers except 2. ∎

Example 7   Determine if y is a function of x and find the domain: $y^2 = x$

Solution   If we let $x = 9$ we have $y^2 = 9$ and $y = \pm 3$. One value of x gives us two values of y. Therefore, y is not a function of x. Since we are squaring the y values, the x values will be non-negative. The domain is $x \geq 0$. ∎

When an equation in two variables is graphed on the coordinate plane, the abcissas are the values of the independent variable and the ordinates are the values of the dependent variable. So the domain of a graphed relation can be determined by finding the horizontal extent of the graph and the range can be determined by finding the vertical extent of the graph.

Two points with the same abscissa and different ordinates would not belong to a function. Two such points would be directly above one another on a vertical line. We can use this fact to determine if a graph is a function by visually determining if any two points on the graph lie on a vertical line. Imagine vertical lines on the graph. If any of these lines intersects the graph more than once, the graph is not a function. This is called the **vertical line test**.

Example 8   Determine if each of the following graphs represents a function.

a.

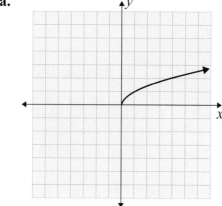

b.

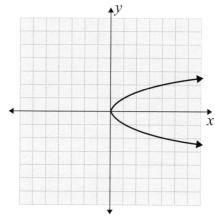

Solution   **a.** This graph would pass the vertical line test since no vertical line will intersect it more than once.

This is the graph of the equation $y = \sqrt{x}$ which was discussed in Example 5.

b. This graph would not pass the vertical line test.

This is the graph of the equation $y^2 = x$ which was discussed in Example 7. ∎

EXERCISE 8-2-1

Determine if the following sets of ordered pairs are functions. Give the domain and range.

1. $\{(0, 3), (-1, 2), (3, -1), (2, 1)\}$ **2.** $\{(-2, 1), (-1, -1), (1, 3), (2, 2)\}$

3. $\{(-2, -2), (-1, 0), (-1, 2), (1, 3)\}$ **4.** $\{(1, 1), (2, 2), (1, -2)\}$

For each equation, determine if y is a function of x. Find the domain.

5. $y = 2x$ **6.** $y = 3x - 1$ **7.** $y = x^2$

8. $y = x^2 + x - 3$ **9.** $y = \dfrac{1}{x-5}$ **10.** $y = \dfrac{1}{2x+1}$

11. $y^2 = x + 1$ **12.** $y = \sqrt{x-1}$

Determine if the following graphs are functions.

13.

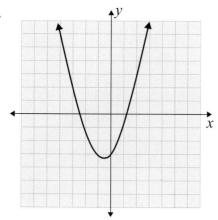

14.

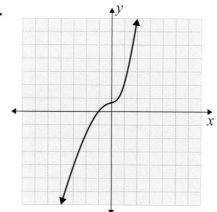

15.

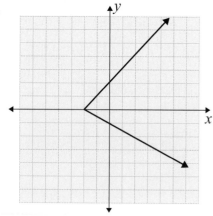

16.

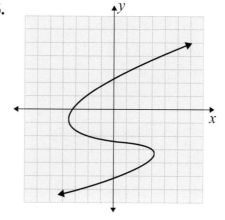

8–3   LINEAR EQUATIONS IN TWO VARIABLES

OBJECTIVES

Upon completion of this section you should be able to:
1. Find the equation of a line given two points on the line.
2. Find the equation of a line given its slope and a point on the line.
3. Write the equation of a line in slope-intercept form.

In Section 8–1 we introduced the equation of a line and learned how to plot its graph. We will now examine the line as a function.

The graph of a line parallel to the y-axis does not represent a function since each x value has many y values. All other lines represent functions. Remember that the graph of a first-degree equation in two variables will form a line.

DEFINITION

The **standard form of the equation of a line** is
$$ax + by = c,$$
where a, b, and c are integers and a and b are not both 0.

Notice that x and y are both to the first power. Any equation that can be written in this form is a linear equation and its graph will be a line.

Example 1   Which of the following equations are linear?

a. $y = 2x$ b. $y = x^2 + x - 3$ c. $y = 3x - 1$

d. $-3x + 2y = 7$ e. $\sqrt{x} - y = 3$ f. $\dfrac{1}{x} + 3y = 6$

g. $x = 4$ h. $y - 5 = 0$

Solution

a. This equation can be rewritten as $2x - y = 0$. It is linear.

b. This equation is not linear because it contains the term x^2.

c. This equation can be rewritten as $3x - y = 1$. It is linear.

d. This equation is linear. It is in standard form.

e. This equation is not linear. The exponent of $\sqrt{x}$ is $\dfrac{1}{2}$, not 1.

f. This equation is not linear. The exponent of $\dfrac{1}{x}$ is -1, not 1.

g. This equation can be rewritten as $x + 0y = 4$. It is linear.

h. This equation can be rewritten as $0x + y = 5$. It is linear. ∎

An important concept related to the equation of a line is that of slope. We can think of the slope of a line as a measure of the steepness and direction of the line. The following definition gives us a more precise meaning.

DEFINITION

The **slope** (m) of a line through the two distinct points (x_1, y_1) and (x_2, y_2) is given by the ratio

$$m = \frac{y_2 - y_1}{x_2 - x_1}, \quad x_1 \neq x_2.$$

Example 2   Find the slope of the line through the points $(1, -5)$ and $(4, 0)$.

Solution   Let $(1, -5)$ be (x_1, y_1) and $(4, 0)$ be (x_2, y_2), and substitute these values into the slope formula.

$$m = \frac{0 - (-5)}{4 - 1} = \frac{5}{3}$$

Notice that if we let $(4, 0)$ be (x_1, y_1) and $(1, -5)$ be (x_2, y_2), we obtain the same value for m.

$$m = \frac{-5 - 0}{1 - 4} = \frac{-5}{-3} = \frac{5}{3} \quad \blacksquare$$

The following sketch shows the slope of the line through $(1, -5)$ and $(4, 0)$.

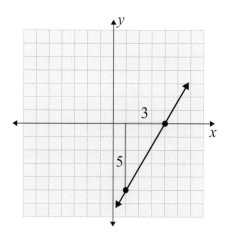

The ratio $\frac{y_2 - y_1}{x_2 - x_1}$ is not dependent on the points chosen as long as they are different points on the line. A non-vertical line can have only one value for its slope. If the line rises from left to right, the slope will be positive. If the line falls from left to right, the slope will be negative. Therefore, the sign of the slope will give the direction of the line. The larger the absolute value of the slope, the steeper the line will be.

Notice that the definition of slope includes the statement $x_1 \neq x_2$. If $x_1 = x_2$, division by zero would result. Consider the line containing the points (3, 5) and (3, 7). This line is vertical. If we try to calculate the slope we have $m = \dfrac{7-5}{3-3} = \dfrac{2}{0}$. Recall that division by zero is undefined and, therefore, the slope of a vertical line (parallel to the y-axis) is undefined.

If a line is horizontal, the y value is the same for each value of x. Consider the line containing the points (3, 5) and (7, 5). This is a horizontal line. Computing the slope gives $m = \dfrac{5-5}{7-3} = \dfrac{0}{4} = 0.$ Thus the slope of a horizontal line (parallel to the x-axis) is zero.

SUMMARY OF SLOPE

If $m > 0$, the line rises from left to right.
If $m < 0$, the line falls from left to right.
If $m = 0$, the line is horizontal.
If m is undefined, the line is vertical.

EXERCISE 8-3-1

Find the slope of the line through the given points. Plot the points and draw the line.

1. (2, 1) and (−3, 11)

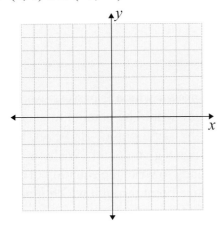

2. (5, 2) and (4, −3)

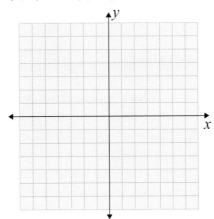

3. (16, 5) and (4, 0)

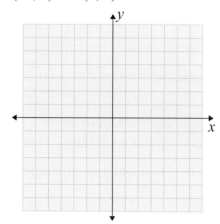

4. (−8, −1) and (3, −5)

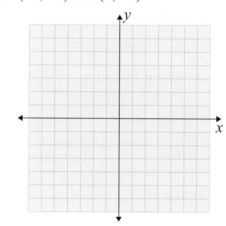

5. (3, 2) and (−4, 2)

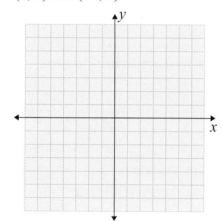

6. (−1, 1) and (−1, 3)

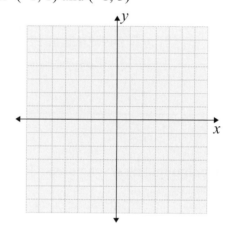

A line is defined by its slope and a point through which it passes. An infinite number of lines could all have the same slope. They would be parallel to each other. An infinite number of lines can all pass through the same point. However, there is only one line that has a given slope and passes through a given point. We have already sketched graphs of linear equations. Now we are ready to write the equation of a line if we are given its graph or if we know its slope and one point through which it passes.

To find a form for the equation of a line with slope m passing through the point (x_1, y_1), first choose some other point on the line and call it (x, y). Then, use the slope formula and the two points (x_1, y_1) and (x, y) to write

$$m = \frac{y - y_1}{x - x_1}.$$

Finally, multiply both sides of this equation by $x - x_1$ to obtain the form

$$m(x - x_1) = y - y_1 \quad \text{or} \quad y - y_1 = m(x - x_1).$$

> **DEFINITION**
>
> The **point-slope form** of the equation of a line is
>
> $$y - y_1 = m(x - x_1).$$

Example 3   Write the standard form of the equation of the line through the point (2, 5) with slope $\dfrac{4}{3}$.

Solution   Since we know a point and the slope, we begin by substituting these values into the point-slope form. Let $(2, 5)$ be (x_1, y_1) and $m = \dfrac{4}{3}$.

$$y - 5 = \frac{4}{3}(x - 2)$$

Then write the equation in standard form.

$$y - 5 = \frac{4}{3}x - \frac{8}{3}$$
$$3y - 15 = 4x - 8$$
$$-4x + 3y = 7 \quad \text{or} \quad 4x - 3y = -7$$

You can check your work by substituting the ordered pair (2, 5) into the equation to see that the equation is true. ■

Example 4   Write the standard form of the equation of the line through the points (3, 5) and (−2, 7).

Solution   We are not given the slope. However, we can find it using the slope formula.

$$m = \frac{7 - 5}{-2 - 3} = \frac{2}{-5} = -\frac{2}{5}$$

Now we can write the equation using the point-slope form. Either point may be used for (x_1, y_1). We will use (3, 5).

$$y - 5 = -\frac{2}{5}(x - 3)$$

Write the equation in standard form.

$$y - 5 = -\frac{2}{5}x + \frac{6}{5}$$
$$5y - 25 = -2x + 6$$
$$2x + 5y = 31$$

Check to see that both points are on this line by substituting into the equation. ■

Recall that a *y*-intercept of an equation is a point where the graph intersects the *y*-axis. It has the form (0, *y*). The **y-intercept of a line** is the point where the line intersects the *y*-axis. The ordered pair representing the *y*-intercept of a line is (0, *b*).

If we know the slope, m, and the y-intercept, $(0, b)$, of a line, then the point-slope form gives us $y - b = m(x - 0)$. Solving for y, we have $y = mx + b$. Notice that the coefficient of x is the slope, m, and the constant term is the ordinate of the y-intercept.

DEFINITION

The **slope-intercept form** of the equation of a line is

$$y = mx + b.$$

Example 5   Write the standard form of the equation of the line with slope $-\dfrac{5}{8}$ and y-intercept $(0, 6)$.

Solution   Since we know the slope and y-intercept, we use the slope-intercept form.

$$y = -\frac{5}{8}x + 6$$

Then write the equation in standard form.

$$5x + 8y = 48 \quad \blacksquare$$

EXERCISE 8-3-2

For each of the following, find the standard form of the equation of the line.

1. $(1, 7)$, $m = 3$

2. $(-2, 0)$, $m = \dfrac{1}{2}$

3. $(-5, 9)$, $m = -8$

4. $(-2, 3)$, undefined slope

5. $(2, 5)$, $m = 0$

6. $(0, 5)$, $m = 2$

7. $(0, -2)$, $m = \dfrac{2}{3}$

8. $(0, 3)$, $m = -3$

Find the standard form of the equation of the line through each of the pairs of points.

9. (2, 1) and (3, 9)   **10.** (−6, 2) and (4,−3)   **11.** (16,−5) and (4, 0)

12. (3, 2) and (−4, 2)   **13.** (−8,−1) and (3, 5)   **14.** (1, −2) and (1, 3)

15. (1, −3) and (0,−3)   **16.** (−2, 3) and (−2, −1)   **17.** $\left(\dfrac{2}{3},\dfrac{1}{2}\right)$ and $\left(\dfrac{-1}{3},\dfrac{5}{2}\right)$

The slope-intercept form of the equation of a line is useful in graphing.

Example 6   Sketch: $y = -\dfrac{5}{8}x + 6$

Solution   We know that the y-intercept is (0, 6). So we plot this point on the graph. The slope is $-\dfrac{5}{8}$ and, since it is a ratio, it does not matter whether we place the negative sign in the numerator or denominator. If we put it in the numerator we obtain $\dfrac{-5}{8}$, which indicates that the change in y is −5 while the change in x is 8. Using these values, we begin at (0, 6) and move 5 units in the negative y direction. We then move 8 units in the positive x direction. The point at which we arrive, (8, 1), is also on the line. Since we now have two points, we may draw the graph of the line as shown.

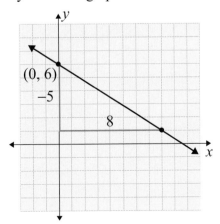

Example 7   Find the slope and the *y*-intercept of the line given by the equation $3x - 2y = 6$. Use this information to sketch the graph.

Solution   To find the slope and *y*-intercept, we use slope-intercept form. Solve for *y*.

$$y = \frac{3}{2}x - 3$$

The slope is $\frac{3}{2}$ and the *y*-intercept is $(0, -3)$. To sketch the graph, we start at the *y*-intercept $(0, -3)$ and move three units in the positive *y* direction and then move two units in the positive *x* direction. The point at which we arrive, $(2, 0)$, is also on the line. Using these two points, we draw the graph.

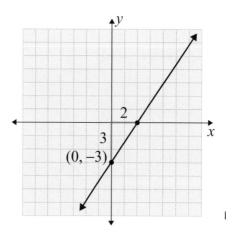

EXERCISE 8-3-3

Give the slope and *y*-intercept and use this information to sketch the graph.

1. $x - 4y = 20$

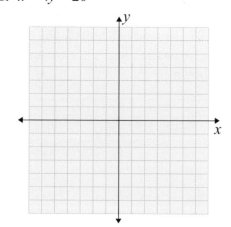

2. $3x + 4y = 8$

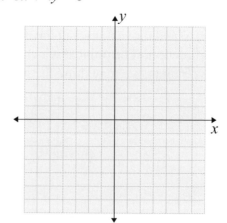

3. $2x - 5y = 15$

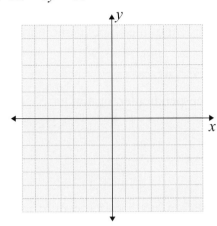

4. $5x - 2y = 0$

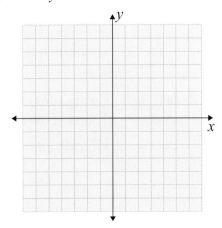

5. $2x + y = 5$

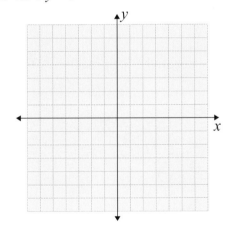

6. $3x + 4y = 0$

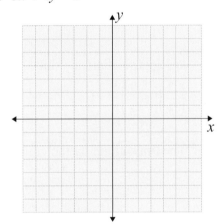

Special properties involving slopes of lines arise when considering equations of parallel and perpendicular lines. Parallel lines are lines in the plane that do not intersect (that is, they do not meet).

> **RULE**
>
> If two distinct lines have the same slope and different *y*-intercepts, they are **parallel**.

Perpendicular lines are lines that form right angles when they intersect. Recall that one number is the negative reciprocal of another if their product is −1.

> **RULE**
>
> If the slope of one line is the negative reciprocal of the slope of another line, the lines are **perpendicular**.

Example 8   Decide if the following pairs of lines are parallel, perpendicular, or neither.

    a. $x - 3y = -15$ and $x - 3y = 21$

    b. $2x - 3y = -12$ and $3x + 2y = 2$

    c. $3x - 5y = 20$ and $3x + 5y = 10$

Solution   To determine the slope of each line, write each equation in slope-intercept form.

    a. The lines given by the equations

$$y = \frac{1}{3}x + 5 \text{ and } y = \frac{1}{3}x - 7$$

        are parallel since their slopes are equal and their y-intercepts are different.

    b. The lines given by the equations

$$y = \frac{2}{3}x + 4 \text{ and } y = -\frac{3}{2}x + 1$$

        are perpendicular since $\dfrac{2}{3}$ and $-\dfrac{3}{2}$ are negative reciprocals.   That is,

$$\left(\frac{2}{3}\right)\left(-\frac{3}{2}\right) = -1.$$

    c. The lines given by the equations

$$y = \frac{3}{5}x - 4 \quad \text{and} \quad y = -\frac{3}{5}x + 2$$

        are neither parallel nor perpendicular. ■

Example 9   Write the equation of **a.** a line parallel and **b.** a line perpendicular to $2x - 3y = -12$, each passing through the point $(-5, 1)$.

Solution   To find the slopes of the parallel and perpendicular lines, write the given equation in slope-intercept form.

$$y = \frac{2}{3}x + 4$$

    a. The line parallel to the given line has the same slope, $\dfrac{2}{3}$. Using point-slope form, we write the equation of this parallel line.

$$y - 1 = \frac{2}{3}(x + 5)$$

$$y - 1 = \frac{2}{3}x + \frac{10}{3}$$

$$3y - 3 = 2x + 10$$

$$-2x + 3y = 13$$

b. The line perpendicular to the given line has a slope that is the negative reciprocal of $\frac{2}{3}$. So $m = -\frac{3}{2}$.

Again using point-slope form, we write the equation of the perpendicular line.

$$y - 1 = -\frac{3}{2}(x + 5)$$

$$y - 1 = -\frac{3}{2}x - \frac{15}{2}$$

$$2y - 2 = -3x - 15$$

$$3x + 2y = -13 \quad \blacksquare$$

EXERCISE 8-3-4

Write the equation, in standard form, of the line passing through the given point and parallel to the given line.

1. $(3, 1)$, $y = 2x + 1$

2. $(5, -1)$, $5x - 2y = -1$

3. $(-8, 1)$, $y = -3$

4. $(4, -9)$, $x = 2$

Write the equation, in standard form, of the line passing through the given point and perpendicular to the given line.

5. $(3, 2)$, $y = \frac{2}{3}x + 1$

6. $(0, -5)$, $x - 2y = 6$

7. $(-1, 4)$, $y = 1$

8. $(2, 5)$, $x = -3$

Practice your skills.

9. A line has an x-intercept of $(-2, 0)$ and a y-intercept of $(0, 7)$. Find the equation of the line in standard form.

10. Find the standard form of the equation of the line passing through the point $(-1, 2)$ and parallel to the y-axis.

11. Find the standard form of the equation of the line passing through the point $(-1, 2)$ and perpendicular to the y-axis.

12. Find the standard form of the equation of the line passing through the point $(1, -2)$ and perpendicular to the x-axis.

13. Find the standard form of the equation of the line with y-intercept $(0, -2)$ and perpendicular to $2x + y = 1$.

14. Find the standard form of the equation of the line with x-intercept $(-2, 0)$ and parallel to $2x + y = 1$.

8–4   SOLVING SYSTEMS OF EQUATIONS IN TWO VARIABLES BY GRAPHING

> **OBJECTIVES**
>
> Upon completion of this section you should be able to:
> 1. Sketch the graphs of two equations in the same coordinate system.
> 2. Solve systems of equations in two variables by graphing.

A **system of equations** in two variables is a set of two or more equations. We know that an equation such as $x + y = 5$ has infinitely many solutions, such as $(3, 2)$, $(6, -1)$, and so on. The equation $2x + y = 8$ also has an infinite number of solutions, such as $(0, 8)$, $(4, 0)$, and so on.

The **solution** to the system

$$\begin{cases} x + y = 5 \\ 2x + y = 8 \end{cases}$$

is the set of all ordered pairs that are solutions to both equations simultaneously. In other words, we are looking for ordered pairs that make both $x + y = 5$ and $2x + y = 8$ true. If we graph both equations on the same coordinate plane, the point or points of intersection will be the solution of the system.

Example 1   Solve by graphing: $\begin{cases} x + y = 5 \\ 2x + y = 8 \end{cases}$

Solution   We recognize that each of these equations is a linear equation. We sketch the graph of each line by finding its x- and y-intercepts or by using its slope and y-intercept.

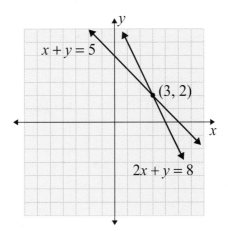

The point of intersection, $(3, 2)$, is the solution of the system and should be labeled as in the illustration. We can check our solution by substituting $(3, 2)$ into both of the equations. ∎

Example 2   Solve by graphing: $\begin{cases} x - y = -1 \\ y = x^2 - 2x + 1 \end{cases}$

Solution   The first equation is a line and can be graphed by finding the intercepts or using slope-intercept form. To graph the second equation, set up a table of values.

$y = x^2 - 2x + 1$

x	−2	−1	0	1	2	3	4
y	9	4	1	0	1	4	9

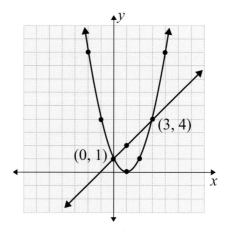

We see that the solutions are (0, 1) and (3, 4). These check in both equations. ■

EXERCISE 8-4-1

Solve the systems by graphing.

1. $\begin{cases} x + y = 2 \\ 2x - y = 1 \end{cases}$

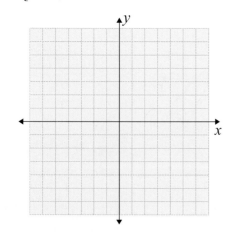

2. $\begin{cases} 3x + y = 0 \\ 2x - y = -5 \end{cases}$

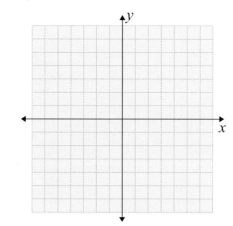

3. $\begin{cases} 3x + y = -9 \\ x - 2y = 4 \end{cases}$

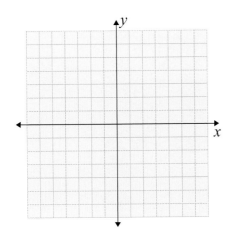

4. $\begin{cases} 3x + y = 8 \\ 2x - y = 7 \end{cases}$

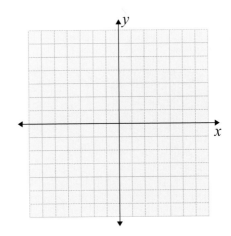

5. $\begin{cases} 3x + 2y = 7 \\ 2x - 3y = -4 \end{cases}$

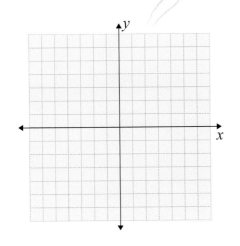

6. $\begin{cases} x - y = -2 \\ y = x^2 \end{cases}$

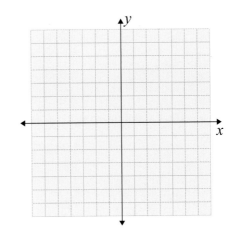

7. $\begin{cases} x + y = -1 \\ y = x^2 + x - 1 \end{cases}$

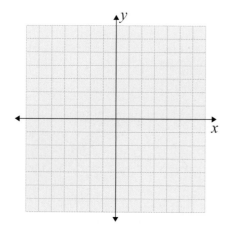

8. $\begin{cases} y = x^2 - 2x + 3 \\ x + y = 5 \end{cases}$

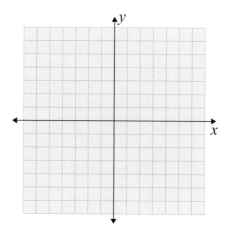

9. Given a system of two linear equations, discuss the possibilities for the solution set.

10. Given a system of one line and one parabola, discuss the possibilities for the solution set.

8–5   THE ALGEBRAIC SOLUTION OF A SYSTEM OF TWO LINEAR EQUATIONS

OBJECTIVES

Upon completion of this section you should be able to:
1. Solve a system of two linear equations by substitution.
2. Solve a system of two linear equations by addition.
3. Solve word problems using a system of two equations in two variables.
4. Classify systems as independent, inconsistent, or dependent.

In Section 8–4 we discussed how to solve systems of equations by graphing the equations to determine the points of intersection. There are two potential problems with this method. First, you must be very accurate in drawing your sketches so that you can read an accurate answer from the graph. Second, the solution points may contain coordinates that are not integers, and therefore the solution may be very difficult to find on the graph. In this section we will discuss two algebraic methods for solving a system of linear equations in two variables.

The first method to be discussed is the *substitution method*.

A PROCEDURE FOR SOLVING A LINEAR SYSTEM BY SUBSTITUTION

1. Solve for one variable in terms of the other in one of the two equations.
2. Substitute this expression into the other equation.
3. Solve the resulting equation.
4. Substitute the value obtained in step 3 into either of the original equations and solve.

Example 1   Solve by the substitution method: $\begin{cases} 2x + 3y = 1 \\ x - 2y = 4 \end{cases}$

Solution   We must solve for one of the unknowns in one of the equations. We can choose either x or y in either the first or second equation. Our choice can be based on obtaining the simplest expression. In this case we will solve for x in the *second* equation, obtaining

$$x = 4 + 2y.$$

Substitute this expression for x into the *first* equation.

$$2(4 + 2y) + 3y = 1$$

Note that we now have an equation in one variable.

Solve for y.

$$2(4+2y)+3y=1$$
$$8+4y+3y=1$$
$$8+7y=1$$
$$7y=-7$$
$$y=-1$$

We can substitute $y=-1$ into either equation to find the corresponding value for x. We will use the second equation, $x-2y=4$.

$$x-2(-1)=4$$
$$x=2$$

When we substitute the ordered pair $(2,-1)$ in both equations, we see that it checks. The solution to the system is $(2,-1)$. ∎

Example 2   Solve by substitution: $\begin{cases} 2x+3y=7 \\ 4x+3y=8 \end{cases}$

Solution   When we solve for x or y in either equation, we will obtain a rational expression, so one choice is as easy as another. We will solve for x in the *first* equation.

$$x=\frac{7-3y}{2}$$

Substitute this expression for x in the *second* equation.

$$4\left(\frac{7-3y}{2}\right)+3y=8$$

Solve.

$$2(7-3y)+3y=8$$
$$14-6y+3y=8$$
$$-3y=-6$$
$$y=2$$

Now substitute $y=2$ into either of the original equations. We will use the first equation, $2x+3y=7$.

$$2x+3(2)=7$$
$$2x=1$$
$$x=\frac{1}{2}$$

Checking, we find that the ordered pair $\left(\frac{1}{2},2\right)$ satisfies both equations and is the solution to the system. ∎

EXERCISE 8-5-1

Solve by the substitution method.

1. $\begin{cases} x+y=3 \\ 2x+y=5 \end{cases}$

2. $\begin{cases} x-2y=5 \\ 2x+y=10 \end{cases}$

3. $\begin{cases} 3x-y=4 \\ x+5y=-4 \end{cases}$

4. $\begin{cases} x+5y=2 \\ 2x+3y=-3 \end{cases}$

5. $\begin{cases} 2x+y=5 \\ 6x+2y=11 \end{cases}$

6. $\begin{cases} 2x-y=1 \\ 5x+2y=-11 \end{cases}$

7. $\begin{cases} x-2y=1 \\ x+4y=22 \end{cases}$

8. $\begin{cases} 2x-y=16 \\ 3x+2y=3 \end{cases}$

9. $\begin{cases} 5x+y=5 \\ 3x-2y=29 \end{cases}$

10. $\begin{cases} x-2y=5 \\ 2x-3y=1 \end{cases}$

The next method we will discuss is the *addition method*. It is based on two rules for solving equations. First, when both sides of an equation are multiplied or divided by a nonzero number, the resulting equation is equivalent to the original equation. Second, when we add the same quantity to both sides of an equation, the resulting equation is equivalent to the original equation.

> ## A PROCEDURE FOR SOLVING SYSTEMS OF LINEAR EQUATIONS BY THE ADDITION METHOD
> 1. Multiply both sides of one or both of the equations by a number or numbers so that the coefficients of one of the variables are equal in absolute value but opposite in sign.
> 2. Add the equations.
> 3. Solve the resulting equation.
> 4. Substitute the value obtained in step 3 into either of the original equations and solve.

Example 3   Solve by addition: $\begin{cases} 2x + y = 5 \\ 3x + 2y = 6 \end{cases}$

Solution   Our goal is to add the two equations and eliminate one of the variables so that we can solve the resulting equation for the other variable. If we add the equations as they are, we will not eliminate a variable. After carefully looking at the problem, we note that the easiest variable to eliminate is y. This is done by multiplying each side of the first equation by -2.

$$\begin{cases} -2(2x + y) = -2(5) \\ 3x + 2y = 6 \end{cases}$$

We obtain the equivalent system.

$$\begin{cases} -4x - 2y = -10 \\ 3x + 2y = 6 \end{cases}$$

Add the equations.

$$\begin{array}{r} -4x - 2y = -10 \\ \underline{3x + 2y = 6} \\ -x = -4 \end{array}$$

Solve the resulting equation. $x = 4$

Find the value of y by substituting $x = 4$ into one of the *original* equations. We will use the first equation.

$$2(4) + y = 5$$
$$y = -3$$

Since the ordered pair $(4, -3)$ makes both equations true, it is the solution of the system. ■

Example 4   Solve by addition: $\begin{cases} 2x+3y=7 \\ 3x+2y=3 \end{cases}$

Solution   We observe that both equations will have to be changed to eliminate one of the variables. If we choose to eliminate x, we can multiply each side of the first equation by 3 and each side of the second equation by -2.

$$\begin{cases} 3(2x+3y)=3(7) \\ -2(3x+2y)=-2(3) \end{cases}$$

We obtain the equivalent system.

$$\begin{cases} 6x+9y=21 \\ -6x-4y=-6 \end{cases}$$

Add the equations and solve the resulting equation.

$$\begin{aligned} 6x+9y &= 21 \\ \underline{-6x-4y} &= \underline{-6} \\ 5y &= 15 \\ y &= 3 \end{aligned}$$

We will substitute $y=3$ into the first equation in our original system to find the value of x.

$$\begin{aligned} 2x+3(3) &= 7 \\ 2x &= -2 \\ x &= -1 \end{aligned}$$

Since the ordered pair $(-1, 3)$ makes both equations true, it is the solution of the system. ■

EXERCISE 8-5-2

Solve by the addition method.

1. $\begin{cases} x+y=2 \\ 4x-y=13 \end{cases}$

2. $\begin{cases} 2x-y=0 \\ x+y=6 \end{cases}$

3. $\begin{cases} 3x - y = -7 \\ x + 2y = 7 \end{cases}$

4. $\begin{cases} 5x - 2y = 54 \\ 2x + 3y = 14 \end{cases}$

5. $\begin{cases} 8x - 3y = 13 \\ 5x - 4y = 6 \end{cases}$

6. $\begin{cases} 2x + y = -3 \\ 7x + 2y = -18 \end{cases}$

7. $\begin{cases} 2x - 5y = 44 \\ 10x + 3y = -4 \end{cases}$

8. $\begin{cases} 2x + 5y = 21 \\ 4x - 3y = -10 \end{cases}$

9. $\begin{cases} 4x + 7y = 27 \\ 3x - 2y = -16 \end{cases}$

10. $\begin{cases} 8x + 3y = -5 \\ 9x + 7y = -2 \end{cases}$

In Chapter 3 we learned how to solve problems involving more than one unknown using just one variable. Now that we have studied systems of equations with two variables, we can choose a variable for each of the unknowns and write a system of equations to solve each problem.

Example 5   Use a system of equations to solve the following word problem:
It takes 3 hours to travel 120 kilometers downstream and 5 hours to make the return trip upstream. Find the average speeds of the boat and the stream.

Solution   Recall that the stream's current adds to the speed of the boat when the boat is traveling downstream. The speed of the stream will also slow the boat when the boat is traveling upstream.

Let b = the speed of the boat.
Let s = the speed of the stream's current.

	r	$\cdot$	t	$=$	d
Downstream	$b+s$		3		120
Upstream	$b-s$		5		120

This gives the system

$$\begin{cases} 3(b+s)=120 \\ 5(b-s)=120. \end{cases}$$

We can solve this system using either the substitution method or the addition method. We will use the addition method.
Divide both sides of the first equation by 3 and both sides of the second equation by 5.

$$\begin{cases} b+s=40 \\ b-s=24 \end{cases}$$

Add the equations.

$$2b=64$$
$$b=32$$

Now substitute this value of b into the first equation.

$$3(32)+3s=120$$
$$96+3s=120$$
$$3s=24$$
$$s=8$$

Thus the speed of the boat is 32 km/hr and the speed of the stream is 8 km/hr. ∎

EXERCISE 8-5-3

Solve the following problems by writing a system of equations and then using the substitution method or the addition method.

1. The total number of students in a class is 48. If there are six more men than women, find the number of men and women in the class.

2. The perimeter of a rectangular lot is 450 feet. If the length is twice the width, find the length and width.

3. The sum of the lengths of the Golden Gate Bridge and the Brooklyn Bridge is 5795 feet. The Golden Gate Bridge is 1010 feet longer than twice the length of the Brooklyn Bridge. Find the length of each bridge.

4. The length of a rectangular lot is eight feet less than three times the width. Find the length and width if the perimeter of the lot is 600 feet.

5. An airliner took 6 hours to travel 3864 kilometers from New York to Las Vegas. The return trip took 4 hours. Find the speeds of the plane and the wind.

6. A professor has 85 students enrolled in two classes, and 49 of these students are freshmen. If two-thirds of the first class and one-half of the second class are freshmen, how many students are in each class?

7. A boat travels 12 kilometers upstream in two hours. The return trip takes one hour. Find the rate of the boat in still water and the rate of the current.

8. A plane travels 1600 kilometers with the wind in two hours. The return trip against the wind takes one-half hour longer. Find the speed of the plane (airspeed) and the speed of the wind.

9. Mike has two more dimes than nickels. The total value of the coins is $2.60. How many of each type of coin does he have?

10. A 40% copper alloy is to be used with a 70% copper alloy to produce 180 kilograms of 60% alloy. How much of each alloy must be used?

A system of two linear equations does not always have a unique solution. Two linear equations will represent one of the following possible situations.

1. Independent equations   The two lines intersect in a single point. In this case we have a unique solution.

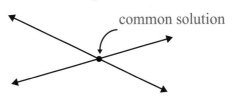
common solution

2. Inconsistent equations   The two lines are parallel. In this case we have no solution.

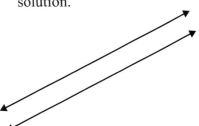

3. Dependent equations   The two equations give the same line. In this case any solution of one equation is a solution of the other.

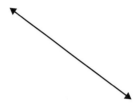

We can recognize any of these three situations by proceeding to work the problem either by the substitution or addition method.

If the equations are **independent**, we will obtain a **unique solution**.
If the equations are **inconsistent**, we will obtain a **contradiction** (a false equation).
If the equations are **dependent**, we will obtain an **identity** (a true equation).

Example 6   Solve: $\begin{cases} x + y = 6 \\ x + y = 8 \end{cases}$

Solution   If we multiply both sides of the second equation by -1 and add, we obtain

$$0 = -2.$$

Since this is a contradiction, the equations are inconsistent and their graphs would be parallel lines. This system has no solution. ∎

Example 7   Solve: $\begin{cases} 2x - 4y = 6 \\ x - 2y = 3 \end{cases}$

Solution   If we multiply both sides of the second equation by -2 and add, we obtain

$$0 = 0.$$

Since this is a true statement, the equations are dependent and their graphs would be the same line. The system has an infinite number of solutions. ∎

EXERCISE 8-5-4

Classify each system as independent, inconsistent, or dependent. If the system is independent, find its solution.

1. $\begin{cases} 2x + y = 1 \\ 3x - y = 9 \end{cases}$

2. $\begin{cases} 2x - y = 1 \\ 6x - 3y = 3 \end{cases}$

3. $\begin{cases} 3x - 2y = 3 \\ 6x - 4y = 1 \end{cases}$

4. $\begin{cases} 2x + y = 5 \\ 10x + 5y = 10 \end{cases}$

5. $\begin{cases} 6x - 4y = 2 \\ 3x - 2y = 1 \end{cases}$

6. $\begin{cases} x - 2y = 15 \\ 3x + y = 3 \end{cases}$

8–6   GRAPHING LINEAR INEQUALITIES

> **OBJECTIVE**
>
> Upon completion of this section you should be able to graph linear inequalities.

In Chapter 2 we constructed line graphs of inequalities.

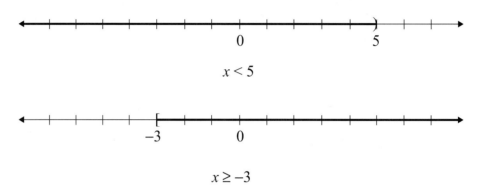

$$x < 5$$

$$x \geq -3$$

These were inequalities involving only one variable. We found that in all such cases the graph was some portion of the number line. Since an equation in two variables gives a graph on the plane, it seems reasonable to assume that an inequality in two variables would graph as some portion or region of the plane. This is in fact the case. The solution of the inequality $x + y < 5$ is the set of all ordered pairs (x, y) such that their sum is less than 5. Note that $x + y < 5$ is a **linear inequality** since $x + y = 5$ is a linear equation.

Example 1   Is each of the following pairs of numbers in the solution set of $x + y < 5$?

$(2, 1), (3, -4), (5, 6), (3, 2), (0, 0), (-1, 4), (-3, 8)$

Solution

$$
\begin{array}{lll}
(2, 1) & x + y < 5 & \\
 & 2 + 1 < 5 & \\
 & 3 < 5 & \text{Yes} \\
(3, -4) & x + y < 5 & \\
 & 3 + (-4) < 5 & \\
 & -1 < 5 & \text{Yes} \\
(5, 6) & x + y < 5 & \\
 & 5 + 6 < 5 & \\
 & 11 < 5 & \text{No} \\
(3, 2) & x + y < 5 & \\
 & 3 + 2 < 5 & \\
 & 5 < 5 & \text{No}
\end{array}
$$

$$(0, 0) \quad x + y < 5$$
$$0 + 0 < 5$$
$$0 < 5 \quad \text{Yes}$$
$$(-1, 4) \quad x + y < 5$$
$$(-1) + 4 < 5$$
$$3 < 5 \quad \text{Yes}$$
$$(-3, 8) \quad x + y < 5$$
$$(-3) + 8 < 5$$
$$5 < 5 \quad \text{No}$$

The line $x + y = 5$ is graphed below. The points from Example 1 are indicated on the graph with answers to the question "Is $x + y < 5$?"

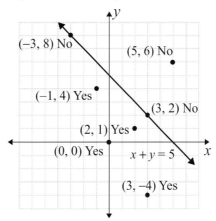

Observe that all "yes" answers lie on the same side of the line $x + y = 5$, and all "no" answers lie on the other side of the line or on the line itself. ∎

The graph of the line $x + y = 5$ divides the plane into three parts: the line itself and the two sides of the line (called **half-planes**).

$$x + y < 5 \text{ is a } \textit{half-plane}$$
$$x + y \leq 5 \text{ is a } \textit{line} \text{ and a } \textit{half-plane}.$$

If one point of a half-plane is in the solution set of a linear inequality, then all points in that half-plane are in the solution set. This gives us a convenient method for graphing linear inequalities.

A PROCEDURE FOR GRAPHING A LINEAR INEQUALITY

To graph a linear inequality:
1. Replace the inequality symbol with an equal sign and graph the resulting line.
2. Check *one* point that is obviously in a particular half-plane of that line to see if it is in the solution set of the inequality.
3. If the point chosen is in the solution set, then that entire half-plane is the solution set. If the point chosen *is not* in the solution set, then the other half-plane is the solution set.

Example 2   Sketch the graph of $2x + 3y > 7$.

Solution   First sketch the graph of the line $2x + 3y = 7$.

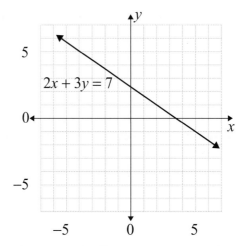

Next choose a point that is *not* on the line $2x + 3y = 7$. If the line does not go through the origin, then the point $(0, 0)$ is always a good choice. Substitute $(0, 0)$ into the inequality $2x + 3y > 7$ to see if it is in the solution set.

$$2(0) + 3(0) > 7$$
$$0 > 7 \quad \text{No}$$

The point $(0, 0)$ is *not* in the solution set. Therefore, the half-plane containing $(0, 0)$ is not the solution set. Hence the other half-plane determined by the line $2x + 3y = 7$ is the solution set.

Since the line itself is not a part of the solution, it is shown as a dashed line and the half-plane is shaded to show the solution set.

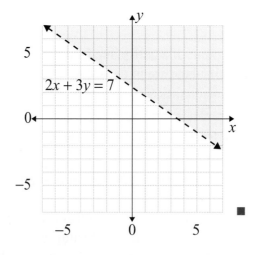

Example 3   Graph the solution for the linear inequality $2x - y \geq 4$.

Solution   First graph $2x - y = 4$.

Since the line graph for $2x - y = 4$ does not go through the origin $(0, 0)$, check that point in the linear inequality.

$$2(0) - (0) \geq 4$$
$$0 \geq 4 \quad \text{No}$$

Since the point $(0, 0)$ is not in the solution set, the half-plane containing $(0, 0)$ is not in the set. Hence the solution is the other half-plane. Notice, however, that the line $2x - y = 4$ is included in the solution set. Therefore, draw a solid line to show that it is part of the graph.

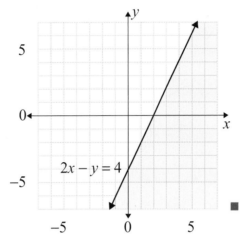

Example 4   Graph $x < y$.

Solution   First graph $x = y$.

Next check a point not on the line. Notice that the graph of the line contains the point $(0, 0)$, so we cannot use it as a check point. To determine which half-plane is the solution set, use any point that is obviously not on the line $x = y$. The point $(-2, 3)$ is such a point.

$$(-2, 3) \quad x < y$$
$$-2 < 3 \quad \text{Yes}$$

Using this information, graph $x < y$.

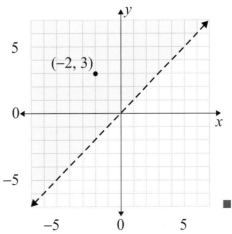

EXERCISE 8-6-1

Graph the linear inequalities.

1. $x + y > 3$

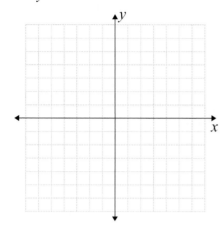

2. $x - y < 5$

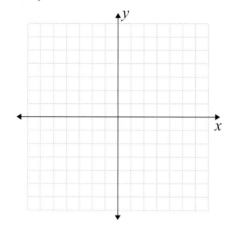

3. $x + 2y < 5$

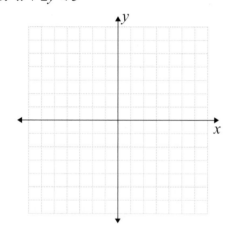

4. $x > y$

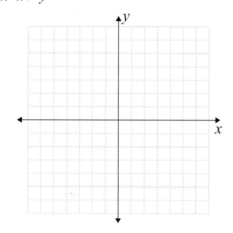

5. $x + 3y \leq 5$

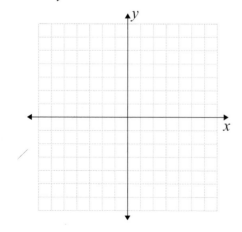

6. $x - 2y \leq 4$

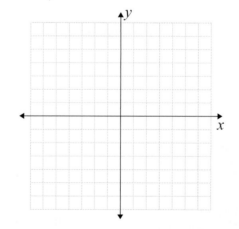

7. $3x - y > 0$

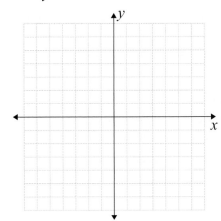

8. $x + 3y > 3$

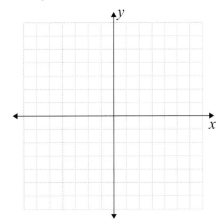

9. $2x + y \leq 3$

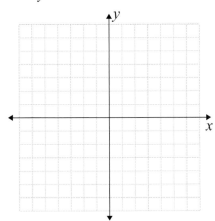

10. $3x - 4y \leq 2$

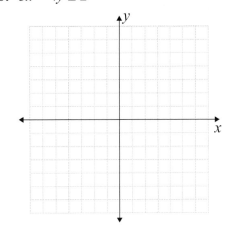

8–7   GRAPHICAL SOLUTION OF A SYSTEM OF LINEAR INEQUALITIES

OBJECTIVES

Upon completion of this section you should be able to:
1. Graph two linear inequalities on the same set of coordinate axes.
2. Determine the region of the plane that is the solution to the system.

You found in Section 8–4 that the solution to a system of equations is the intersection of the solutions to each of the equations. In the same manner, the solution to a system of linear inequalities is the intersection of the half-planes (and perhaps lines) that are solutions to each individual linear inequality. In other words, $x + y > 5$ has a solution set that is a half-plane, and $2x - y < 4$ has a solution set that is a half-plane. Therefore, the system

$$\begin{cases} x + y > 5 \\ 2x - y < 4 \end{cases}$$

has as its solution set the region of the plane that is the intersection of the two half-planes.

To graph the solution to this system we graph each linear inequality on the same set of coordinate axes and indicate the intersection of the two solution sets.

Example 1   Graph the solution: $\begin{cases} x + y > 5 \\ 2x - y < 4 \end{cases}$

Solution   First graph the lines $x + y = 5$ and $2x - y = 4$.

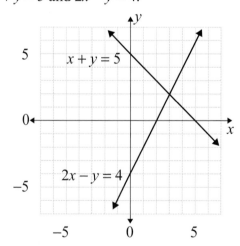

Checking the point $(0, 0)$ in the inequality $x + y > 5$ indicates that the point $(0, 0)$ is not in its solution set. We indicate the solution set of $x + y > 5$ with yellow shading.

Checking the point $(0, 0)$ in the inequality $2x - y < 4$ indicates that the point $(0, 0)$ is in its solution set. We indicate the solution set of $2x - y < 4$ with blue shading.

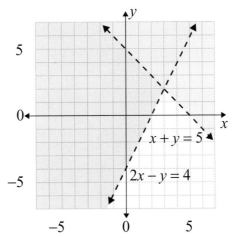

The intersection of the two solution sets is that region of the plane in which the two shadings intersect. This region is shown in green in the graph.

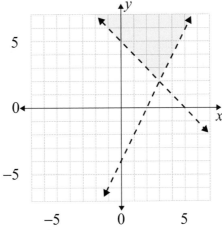

The results indicate that all points in the shaded section of the graph would be in the solution sets of $x + y > 5$ *and* $2x - y < 4$ at the same time. Note that the solution does not include the lines. ■

Example 2   Graph the solution: $\begin{cases} x + y \geq 5 \\ 2x - y < 4 \end{cases}$

Solution   This system is similar to the system in Example 1. However, the solution includes points *on* the line $x + y = 5$ which also satisfy the second inequality. We indicate the solution as follows.

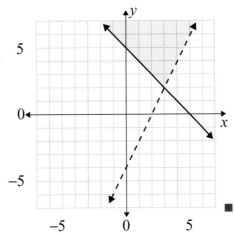

EXERCISE 8-7-1

Solve the systems by graphing.

1. $\begin{cases} x + y > 2 \\ 2x - y < 1 \end{cases}$

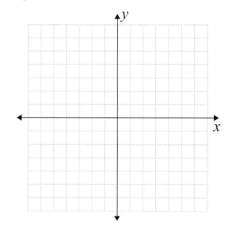

2. $\begin{cases} 3x + y > 6 \\ x - 2y < 4 \end{cases}$

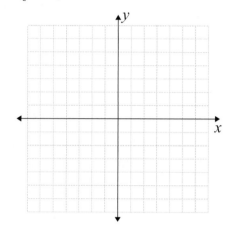

3. $\begin{cases} x + y > 0 \\ x - 3y > 3 \end{cases}$

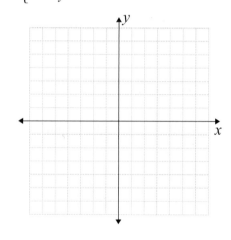

4. $\begin{cases} x + 2y \le 4 \\ 2x - y \ge 6 \end{cases}$

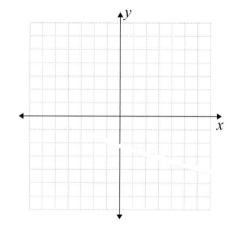

5. $\begin{cases} 4x - y < 4 \\ x + 2y \le 2 \end{cases}$

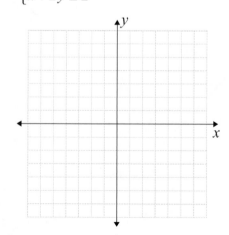

6. $\begin{cases} x - 2y < 3 \\ 2x + y > 8 \end{cases}$

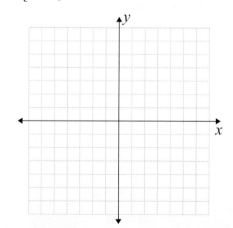

7. $\begin{cases} 4x - 3y \le -12 \\ x + 4y > 6 \end{cases}$

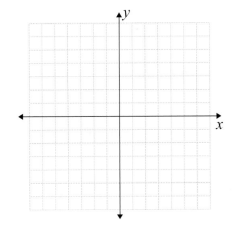

8. $\begin{cases} x - y \le 0 \\ x + y \ge 0 \end{cases}$

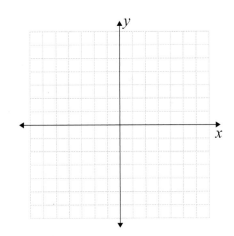

9. $\begin{cases} 3x + y \ge -9 \\ x - 2y \ge 4 \end{cases}$

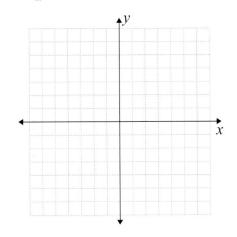

10. $\begin{cases} x - 3y < 5 \\ 2x + 3y \le 8 \end{cases}$

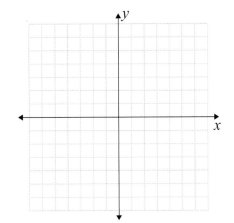

8–8   THE ALGEBRAIC SOLUTION OF THREE EQUATIONS WITH THREE VARIABLES

OBJECTIVES

Upon completion of this section you should be able to:
1. Solve a system of three equations in three variables.
2. Solve word problems using a system of three equations in three variables.

A first-degree polynomial equation in three variables is the equation of a plane in three-dimensional space. Three planes in space might intersect in a single point. However, there are other possibilities, such as all three planes being parallel. In this section we are interested only in planes that intersect in a single point (that is, systems that have a unique solution), and in an algebraic method of finding the solution.

A point in three-dimensional space is represented by an **ordered triple** of numbers. We use x, y, and z as the three variables and the ordered triple is (x, y, z). The ordered triple $(2, 3, -1)$ represents a point such that $x = 2$, $y = 3$, and $z = -1$.

The method we will use reduces a system of three equations in three variables to two equations in two variables by addition. This method is best illustrated by example.

Example 1   Solve the system: $\begin{cases} 2x + 3y - z = 11 & (1) \\ x + 2y + z = 12 & (2) \\ 3x - y + 2z = 5 & (3) \end{cases}$

Solution   **Step 1**   Choose any two equations and eliminate any one of the three variables by addition.

Note that this gives a wide range of choices. We can choose on the basis of the variable that is easiest to eliminate. In this case we will choose to eliminate z by adding equations (1) and (2).

Equation (1) added to equation (2) yields

$$3x + 5y = 23.$$

Step 2   Choose another pair of equations and eliminate the *same* variable by addition.

The choices here are not as broad because we must eliminate the same variable as in step 1 but cannot use the same two equations as before. In this example we must eliminate z and can use equations (1) and (3) or equations (2) and (3). We will choose equations (1) and (3) and eliminate z by multiplying both sides of equation (1) by 2 and adding the result to equation (3).

Twice equation (1) added to equation (3) yields

$$7x + 5y = 27.$$

Step 3   Solve the system of two equations with two variables that results from steps 1 and 2.

$$\begin{cases} 3x + 5y = 23 \\ 7x + 5y = 27 \end{cases}$$

Using either substitution or addition to solve this system, we find

$$x = 1$$
$$y = 4.$$

Step 4   Use the values from step 3 in any one of the original equations to find the other variable.

Using equation (2) and $x = 1$, $y = 4$, we have

$$x + 2y + z = 12$$
$$1 + 2(4) + z = 12$$
$$z = 3.$$

Step 5   Check the solution in all three of the original equations to see that each equation is satisfied.

In our example the solution $(1, 4, 3)$ checks in all three equations. ∎

Example 2   Solve the system: $\begin{cases} x - y + 2z = 6 & (1) \\ 2x + 3y - z = -3 & (2) \\ 3x + 2y + 2z = 5 & (3) \end{cases}$

Solution   **Step 1**   Eliminate y by multiplying both sides of equation (1) by 3 and adding the result to equation (2).

Three times equation (1) added to equation (2) gives

$$5x + 5z = 15.$$

Step 2   Eliminate y by multiplying both sides of equation (1) by 2 and adding the result to equation (3). This yields

$$5x + 6z = 17.$$

Step 3   Solve the system: $\begin{cases} 5x + 5z = 15 \\ 5x + 6z = 17 \end{cases}$

We find the solution to be

$$x = 1$$
$$z = 2.$$

Step 4   Substitute $x = 1$ and $z = 2$ in equation (1) to solve for y. $y = -1$

Step 5   The ordered triple $(1, -1, 2)$ checks in all three equations and is the solution to the system. ■

If one of the variables is missing from one or more of the equations, finding the solution becomes easier.

Example 3   Solve the system:
$$\begin{cases} x + z = 3 & (1) \\ 2x + y + z = 3 & (2) \\ 3x - y + 2z = 8 & (3) \end{cases}$$

Solution   Note that equation (1) contains only two of the variables. We can eliminate y using equations (2) and (3), and then we will have two equations in two variables (x and z).

Equation (2) added to equation (3) gives

$$5x + 3z = 11.$$

Now we solve the system

$$\begin{cases} x + z = 3 \\ 5x + 3z = 11. \end{cases}$$

We obtain as our solution

$$x = 1$$
$$z = 2.$$

Substituting these values into equation (2) yields $y = -1$.

The solution $(1, -1, 2)$ checks in all three equations. ■

Example 4   Jena has $5.10 in nickels, dimes, and quarters. She has a total of 37 coins. The number of nickels and quarters combined is three more than the number of dimes. Find the number of each kind of coin Jena has.

Solution   To solve this problem, we will use three variables and write three equations.

Let n = number of nickels.
Let d = number of dimes.
Let q = number of quarters.

Since the value of all the coins is $5.10, we may write the first equation as
$$0.05n + 0.10d + 0.25q = 5.10 \quad \text{or} \quad 5n + 10d + 25q = 510.$$

Either equation can be rewritten as

$$n + 2d + 5q = 102.$$

Since there are a total of 37 coins, we can write the second equation as

$$n + d + q = 37.$$

The third equation is obtained from the statement that the number of nickels and quarters combined is three more than the number of dimes.

$$n + q = d + 3$$

$$\text{or} \quad n - d + q = 3$$

We now have the system

$$\begin{cases} n + 2d + 5q = 102 \\ n + d + q = 37 \\ n - d + q = 3. \end{cases}$$

Solving, we obtain

$$n = 8$$
$$d = 17$$
$$q = 12. \quad \blacksquare$$

EXERCISE 8-8-1

Solve.

1. $\begin{cases} x + y + z = 6 \\ 2x - y + z = 3 \\ x - y + 2z = 5 \end{cases}$

2. $\begin{cases} x + 2y + z = 0 \\ x - 3y - z = -2 \\ x + y - z = -2 \end{cases}$

3. $\begin{cases} 2x + y + z = 0 \\ 3x - 2y - z = -11 \\ x - y + 2z = 3 \end{cases}$

4. $\begin{cases} x - y + z = 8 \\ 5x + 4y - z = 7 \\ 2x + y - 3z = -7 \end{cases}$

5. $\begin{cases} x + 5y - 2z = 13 \\ 6x + y + 3z = 4 \\ x - y + 2z = -5 \end{cases}$

6. $\begin{cases} x + y = 6 \\ 2x - y + z = 7 \\ x + y - 3z = 12 \end{cases}$

7. $\begin{cases} 3x + 4y + z = -2 \\ y + z = 1 \\ 2x - y - z = -5 \end{cases}$

8. $\begin{cases} y - z = -3 \\ x + y = 1 \\ 2x + 3y + z = 1 \end{cases}$

9. Philip has $4.50 in nickels, dimes, and quarters. He has a total of 28 coins. The number of dimes is twice the number of nickels. How many of each type of coin does he have?

10. The sum of three numbers is 58. Twice the first number added to the sum of the second and third numbers is 71. If the first number is added to four times the second number and the sum is decreased by three times the third number, the result is 18. Find the numbers.

11. A chemist wishes to make 9 liters of a 30% acid solution by mixing three solutions of 5%, 20%, and 50%. How much of each solution must the chemist use if twice as much 50% solution is used as 5% solution?

12. Kristy buys 11 rolls of three different kinds of wallpaper, some at $8.00 a roll, some at $7.00 a roll, and some at $5.00 a roll. She has twice as many rolls of $5.00 paper as she does of $7.00 paper. If the total bill for the wallpaper is $67.00, how many rolls of $5.00 paper did she buy?

CHAPTER 8 SUMMARY

The number in brackets refers to the section of the chapter that discusses the concept.

Terminology

- A **relation** is a set of ordered pairs. [8–1]
- The **Cartesian coordinate system** is a method of uniquely naming each point in the plane. [8–1]
- A **linear equation** is a first-degree equation in two variables. [8–1]
- A **y-intercept** of a graph is a point where the graph intersects the y-axis. [8–1]
- An **x-intercept** of a graph is a point where the graph intersects the x-axis. [8–1]
- The **domain** is the set of all first numbers (**abscissas**) in the ordered pairs of a relation. [8–2]
- The **range** is the set of all second numbers (**ordinates**) in the ordered pairs of a relation. [8–2]
- A **function** is a relation such that each element of the domain is related to exactly one element in the range. [8–2]
- The **standard form** of the equation of a line is $ax + by = c$, where a, b, and c are integers and a and b are not both 0. [8–3]
- The **slope** (m) of a line is given by $m = \dfrac{y_2 - y_1}{x_2 - x_1}$, $x_1 \neq x_2$. [8–3]
- The **point-slope form** of the equation of a line is $y - y_1 = m(x - x_1)$. [8–3]
- The **slope-intercept form** of the equation of a line is $y = mx + b$, where m is the slope of the line and $(0, b)$ is the y-intercept. [8–3]
- **Parallel** lines have the same slope. [8–3]
- **Perpendicular** lines have slopes that are negative reciprocals. [8–3]
- A **system of equations** is a set of two or more equations. [8–4]

- The **solution** of a system of equations in two variables is the set of all ordered pairs that are solutions to all equations of the system. [8–4]
- A **half-plane** is that portion of the plane that lies on one side of a line. [8–6]
- A **linear inequality** is a first-degree inequality in two variables. [8–6]
- A **system of inequalities** is a set of two or more inequalities. [8–7]

Rules and Procedures

Graphing

- To graph an equation in two variables in the Cartesian coordinate system, find ordered pairs that are solutions to the equation and represent these as points on the coordinate system. Find the pattern and complete the graph of the equation. [8–1]
- To graph a linear inequality: [8–6]
  1. Replace the inequality symbol with an equal sign and graph the resulting line.
  2. Check *one* point that is obviously in a particular half-plane of that line to see if it is in the solution set of the inequality.
  3. If the point chosen *is* in the solution set, then that entire half-plane is the solution set. If the point chosen is *not* in the solution set, then the other half-plane is the solution set.

Solving Systems of Equations and Inequalities

- To solve a system of equations in two variables by graphing, graph the equations on the same coordinate plane. The point(s) of intersection will be the solution(s) of the system. [8–4]

- To solve a system of linear equations in two variables using the substitution method: [8–5]
    1. Solve for one variable in terms of the other in one of the two equations.
    2. Substitute this expression into the other equation.
    3. Solve the resulting equation.
    4. Substitute the value obtained in step 3 into either of the original equations and solve.
- To solve a system of linear equations in two variables using the addition method: [8–5]
    1. Multiply both sides of one or both of the equations by a number or numbers so that the coefficients of one of the variables are equal in absolute value but opposite in sign.
    2. Add the equations.
    3. Solve the resulting equation.
    4. Substitute the value obtained in step 3 into either of the original equations and solve.
- To solve a system of linear inequalities by graphing, determine the region of the plane that satisfies both inequalities. [8–7]
- The algebraic solution of three equations with three variables involves choosing any two of the three equations and then eliminating one of the variables by addition. Next, choose any other two of the three equations and eliminate the same variable. The two equations with two variables thus obtained are solved as a system of two linear equations and the solution is substituted into any one of the three original equations to obtain the value of the third variable. The numbers obtained must be a solution of each of the three equations to be a solution of the system. [8–8]

Chapter 8 Review

Complete the table of values and sketch the graph.

1. $y = x - 3$

x	-2	-1	0	1	2
y					

2. $y = x^2 - 1$

x	-2	-1	0	1	2
y					

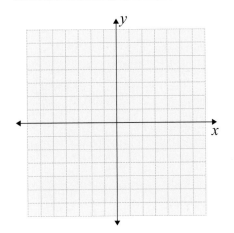

Decide if y is a function of x. If the equation represents a function, find its domain.

3. $y = x^2 + 1$

4. $x = y^2 - 2$

5. $y = \dfrac{1}{x^2 - 1}$

Find the slope of the lines through the given points.

6. $(-1, 3)$ and $(1, 2)$

7. $(1, 5)$ and $(1, -6)$

8. $(2, -4)$ and $(-3, -4)$

9. Find the standard form of the equation of the line through the points (4, −6) and (−3, 5).

10. Sketch the graph of the line through the point (2, 3) with slope $m = \dfrac{1}{2}$.

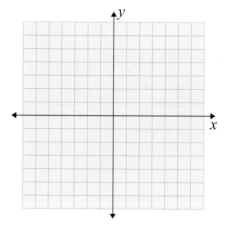

11. Find the slope and y-intercept of $2x - 5y = 10$.

12. Find the standard form of the equation of the line that has a slope of $\dfrac{1}{2}$ and a y-intercept of 3.

13. Find the standard form of the equation of the line that is parallel to the line $3x + 8y = 1$ and passes through the point (−2, 7).

14. Find the standard form of the equation of the line passing through the point (1, −5) with undefined slope.

Solve by graphing.

15. $\begin{cases} x + y = -1 \\ y = x^2 - 1 \end{cases}$

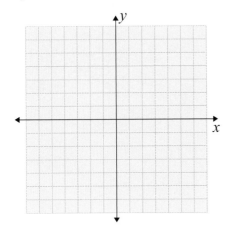

Solve algebraically.

16. $\begin{cases} x + y = -1 \\ 2x - y = 4 \end{cases}$

17. $\begin{cases} 2x + y = -3 \\ x - 2y = -4 \end{cases}$

18. $\begin{cases} 3x - y = 7 \\ 2x + y = 8 \end{cases}$

19. A boat travels 23 kilometers per hour downstream and 15 kilometers per hour upstream. Find the speed of the current and the speed of the boat.

Classify each of the following systems as independent, inconsistent, or dependent. If the system is independent, find its solution.

20. $\begin{cases} x + y = 1 \\ 2x - y = 5 \end{cases}$ **21.** $\begin{cases} 2x - y = 4 \\ 6x - 3y = 8 \end{cases}$ **22.** $\begin{cases} 3x + 6y = 15 \\ 2x + 4y = 10 \end{cases}$

Graph the linear inequalities.

23. $x + y < 0$ **24.** $x + 2y \geq 3$

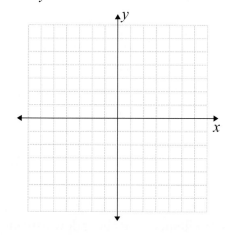   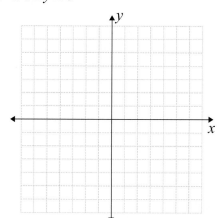

Solve the systems by graphing.

25. $\begin{cases} 3x + y > 0 \\ 2x - y < -4 \end{cases}$

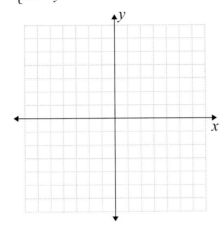

26. $\begin{cases} 2x - y \geq 2 \\ x + y \leq 3 \end{cases}$

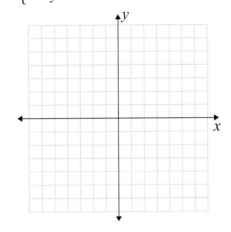

Solve the systems.

27. $\begin{cases} x + y - z = 4 \\ 2x + y + z = 3 \\ 2x + 2y + z = 5 \end{cases}$

28. $\begin{cases} x + y - z = -3 \\ x + y + z = 3 \\ 3x - y + z = 7 \end{cases}$

29. Carly buys 11 liters of three different kinds of paint, some at $5.00 a liter, some at $4.00 a liter, and some at $3.00 a liter. She has twice as many liters of $3.00 paint as he does of $4.00 paint. If her total bill for the paint is $40.00, how many liters of $3.00 paint did she buy?

CHAPTER 8 PRACTICE TEST

1. In the equation $y = \sqrt{x+2}$, is y a function of x?   **1.** _____
    If so, what is the domain?

2. Given the function $y = x^2 - 2$:
    a. Complete the table of values.   **2. a.**

x	-2	-1	0	1	2
y					

    b. Graph.   **2. b.**

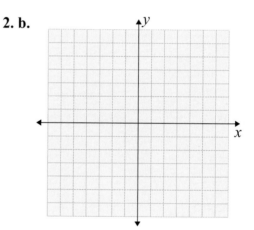

3. Write the standard form of the equation   **3.** _____
    of the line passing through the points
    $(0, -2)$ and $(5, 1)$.

4. Find the slope and y-intercept for $2x - 3y = 1$.   **4.** _____

5. Find the equation of the line perpendicular to $5x + y = 1$ and passing through the point $(0, -2)$.

5.

6. Solve by graphing: $\begin{cases} y - x = 1 \\ y = x^2 - 1 \end{cases}$

6.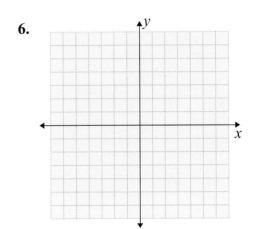

7. Solve by graphing: $\begin{cases} x + y = 4 \\ 2x - y = 5 \end{cases}$

7.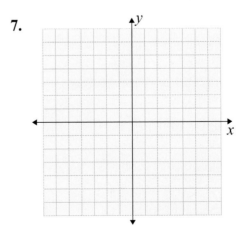

Solve algebraically.

8. $\begin{cases} 2x + y = 2 \\ x - 3y = 15 \end{cases}$

8. _____

9. $\begin{cases} 2x + 3y = -2 \\ 3x + 2y = 7 \end{cases}$

9. _____

10. Sketch the graph of the linear inequality:
$x + y < 6$

10.

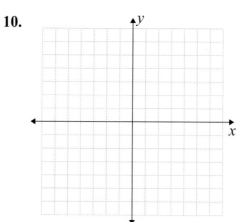

11. Solve by graphing:

$$\begin{cases} x + y \le -4 \\ 5x - 3y \ge 10 \end{cases}$$

11.

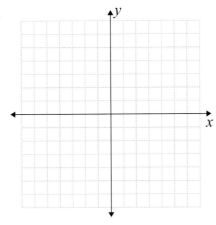

12. Solve:

$$\begin{cases} x + 2y + z = 1 \\ 2x + 3y - z = 0 \\ x - 2y + 3z = 7 \end{cases}$$

12. _____

13. The sum of three numbers is zero. Twice the first number added to the second is 11 less than the third. The third number is 13 more than the second. Find the numbers.

13. _____

END OF BOOK TEST

1. Evaluate: $-4 - 5 - (-7)$

   1. _____

2. Simplify: $\dfrac{24a^4}{-3a^2}$

   2. _____

3. Multiply: $2a^2b^3\left(2a^3b - 3b\right)$

   3. _____

4. Evaluate: $7 - 3\left[3 + (4-1)^2 - 2\right]$

   4. _____

5. Simplify: $5x^2 - \left[7x - 3x(4x+1)\right]$

   5. _____

Solve for x.

6. $3(x+2) - 3 = 5(x-1)$

   6. _____

7. $\dfrac{3x}{2} + \dfrac{x}{4} = 7$

   7. _____

8. $\frac{1}{3}(2x - 3a) = 2(x + a)$　　　　8. _____

9. $-2x \leq -6$　　　　9. _____

10. $|x - 3| > 7$　　　　10. _____

Factor completely.

11. $9a^2b^2 - 6ab^3 - 12a^3b^2$　　　　11. _____

12. $8x^3 + 125$　　　　12. _____

13. $2ax - 20a - 3x + 30$　　　　13. _____

14. $8p^2 + 20p - 12$　　　　14. _____

Simplify.

15. $\dfrac{x^2+5x-14}{x^2-3x+2} \cdot \dfrac{x^2-6x+5}{x^2-25}$

15. _____

16. $\dfrac{6y+3}{4y+y^2} \div \dfrac{2y^2-5y-3}{3y-y^2}$

16. _____

17. $\dfrac{5}{x} + \dfrac{4}{x+3}$

17. _____

18. $\dfrac{5m+2}{m^2+4m-5} - \dfrac{2m}{m-1}$

18. _____

19. $\dfrac{1-\dfrac{3}{x}}{x^2-9}$

19. _____

20. Solve: $\dfrac{5}{x^2+4x-21} - \dfrac{2}{x+7} = \dfrac{3}{x-3}$

20. _____

21. Evaluate: $16^{-\frac{3}{2}}$

21. _____

22. Evaluate: $\sqrt[3]{-27}$

22. _____

Simplify.

23. $\sqrt{27x^4}$

23. _____

24. $5\sqrt{7} - \sqrt{28}$

24. _____

25. $\left(2\sqrt{3} + \sqrt{2}\right)\left(3\sqrt{3} - \sqrt{2}\right)$

25. _____

26. $\dfrac{x-4}{\sqrt{x}-2}$

26. _____

Simplify. Write final answers without negative exponents.

27. $\left(-3xy^4\right)^3$

27. _____

28. $\left(3a^2b\right)^2\left(-2ab^3\right)^3$

28. _____

29. $\left(\dfrac{3x^{-4}}{5y^{-2}}\right)^{-2}$ 29. _____

30. $a^{\frac{2}{3}}a^{\frac{-1}{6}}$ 30. _____

31. $\left(\dfrac{9x^{-2}}{16y^4}\right)^{\frac{1}{2}}$ 31. _____

32. $(1+2i)-3(-2-5i)$ 32. _____

Solve for real or complex roots.

33. $3z^2-4=0$ 33. _____

34. $6x^2-27x-15=0$ 34. _____

35. $\dfrac{2t}{7}-2+\dfrac{7}{2t}=0$ 35. _____

36. $\sqrt{x+2} = x$

36. _____

37. Write the equation, in standard form, of the line passing through the points $(7, 0)$ and $(3, -5)$.

37. _____

38. Solve by graphing: $\begin{cases} 3x + y = 8 \\ 2x - y = 7 \end{cases}$

38. _____

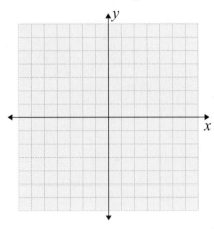

39. Solve: $\begin{cases} 2x + 3y = 6 \\ 3x - 2y = -17 \end{cases}$

39. _____

40. Solve by graphing: $\begin{cases} x + y \le 0 \\ 2x - y > 4 \end{cases}$

40. _____

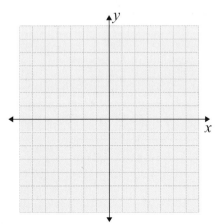

41. Solve: $\begin{cases} 4x - 2y + z = 4 \\ 2x - y + 3z = -3 \\ x + y - 2z = 7 \end{cases}$

41. _____

42. During a week of sales, Amanda earned $23 more commission than Burton. Carlos earned twice as much as Burton. Find the commission for each salesperson if their total combined commission was $3555.

42. _____

43. Christian drove his boat downstream 48 miles and then back. The entire trip took five hours. If the speed of the current is 4 miles per hour, find the rate of Christian's boat in still water.

43. _____

44. Eric can mow the lawn in 6 hours. Eric and Stephanie, working together, can mow the lawn 2 hours and 24 minutes. How long will it take Stephanie to do the job alone?

44. _____

45. Mary has 80 feet of fence to enclose a rectangular garden. If she wants the garden to have an area of 300 square feet, what should the dimensions be?

45. _____

ANSWER KEY

Chapter 1 Survey

1. irrational [1–1] **2.** 31 [1–1] **3.** 4 [1–1] **4.** −47 [1–1] **5.** −6 [1–2] **6.** −31 [1–2] **7.** 42 [1–3] **8.** −6 [1–3] **9.** −30 [1–3]

10. x^7 [1–4] **11.** $-35a^5b^5$ [1–4] **12.** $-\dfrac{9x^6}{y^2}$ [1–4] **13.** $x^2y - 5xy + 7xy^2$ [1–5] **14.** $-3a^3 + 6a^2 - 3a$ [1–6] **15.** 10 [1–7]

16. $3x^2 + 4xy$ [1–7] **17.** 22 [1–8]

Exercise 1-1-1

1. integer, rational **3.** rational **5.** counting, integer, rational, whole **7.** irrational

Exercise 1-1-2

1. −5 **3.** 2 **5.** $1\dfrac{1}{2}$ **7.** $7 + 2$ **9.** $-\pi$ **11.** $-x$

Exercise 1-1-3

1. 2 **3.** 5 **5.** 0 **7.** π

Exercise 1-1-4

1. 17 **3.** −15 **5.** 22 **7.** −4 **9.** 0 **11.** 40 **13.** $-\dfrac{2}{7}$ **15.** 0 **17.** $\dfrac{1}{6}$ **19.** 6 degrees **21.** a gain of 4 points **23.** a loss of 4 yards

Exercise 1-2-1

1. 6 **3.** −8 **5.** 10 **7.** 17 **9.** −23 **11.** −11 **13.** −27 **15.** 0 **17.** $-\dfrac{8}{9}$ **19.** $-\dfrac{5}{8}$

Exercise 1-2-2

1. 15 **3.** −25 **5.** 3 **7.** −15 **9.** $-\dfrac{2}{5}$ **11.** −3 **13.** 14 **15.** $\dfrac{5}{8}$ **17.** 8 **19.** 40 **21.** 56° **23.** loss: 8 pounds

Exercise 1-3-1

1. 12 **3.** −10 **5.** 12 **7.** 1 **9.** $-\dfrac{14}{15}$ **11.** 5 **13.** −4 **15.** 0 **17.** −4 **19.** undefined

Exercise 1-3-2

1. −70 **3.** −80 **5.** 48 **7.** −108 **9.** −1 **11.** 0 **13.** 264 **15.** $\dfrac{7}{10}$ **17.** −80 **19.** −37 **21.** −5 **23.** $-\dfrac{9}{5}$ **25.** $\dfrac{1}{4}$ **27.** 11

Exercise 1-4-1

1. 9 **3.** 9 **5.** −9 **7.** −8 **9.** x^8 **11.** 3^7 **13.** $(-5)^4$ **15.** x^8 **17.** x^4y^3

Exercise 1-4-2

1. $10x^5$ **3.** $-21x^5$ **5.** $2a^2b^3$ **7.** $-6x^3y^4$ **9.** $-\dfrac{3}{5}x^{11}$ **11.** $-48x^3y^6$ **13.** $\dfrac{10}{7}x^3y^9z^3$ **15.** $60a^2b^3c^5d^2$

Exercise 1-4-3

1. x **3.** $\dfrac{1}{x^5}$ **5.** $-\dfrac{1}{5}$ **7.** $-\dfrac{2x^5}{y^3}$ **9.** $\dfrac{1}{4}$ **11.** x^5y^3 **13.** $-\dfrac{4y^4}{x^3}$ **15.** 1 **17.** $-\dfrac{5x}{7z}$ **19.** $-\dfrac{3}{4}wx^2$

Exercise 1-5-1

1. $12a$ **3.** $-7a$ **5.** $-4xy$ **7.** in simplest form **9.** in simplest form **11.** 0 **13.** $-15a^2b+5ab^2$ **15.** in simplest form **17.** $-3xy+8a$
19. $4xyz+15xy+7xy^2$

Exercise 1-6-1

1. $3x+6y$ **3.** x^2+4x **5.** $-2a+7b$ **7.** $10x+5y-15z$ **9.** $-2x^2+4xy$ **11.** $-3m-5n+p$ **13.** $-7xy+14y^2-21y$
15. $6x+15x^2-12x^3$ **17.** $6x^3y^2-4x^2y^3+2x^2y^2$ **19.** $2x^3y^2z-10xy^3z^2+18x^2yz^2$

Exercise 1-7-1

1. 18 **3.** 4 **5.** 4 **7.** 4 **9.** -3 **11.** 20 **13.** 10 **15.** 21 **17.** $-\dfrac{1}{2}$ **19.** $-\dfrac{1}{8}$ **21.** 3 **23.** 26 **25.** 0 **27.** -18 **29.** $\dfrac{1}{10}$

Exercise 1-7-2

1. -16 **3.** -53 **5.** $3x-1$ **7.** $2x+1$ **9.** $-6x+5$ **11.** $11x+9$ **13.** $2x-3$ **15.** $2a-4$ **17.** $12x-18$ **19.** $11a-5$ **21.** $5x+4$
23. $-13a+24$ **25.** $16a-9b$

Exercise 1-8-1

1. -16 **3.** 25 **5.** 25 **7.** 400 **9.** 14 **11.** -36 **13.** 38 cm **15.** 32 m **17.** 42.9 ft **19.** 220 miles **21.** 44 m **23.** $20°$ **25.** 18 cm^3
27. 462 ft^3 **29.** 462 m^3 **31.** 3340 cm

Chapter 1 Review

1. integer, rational [1–1] **2.** rational [1–1] **3.** counting, whole, integer, rational [1–1] **4.** $\sqrt{3}$ [1–1] **5.** -2 [1–1] **6.** -22 [1–1]
7. $\dfrac{2}{9}$ [1–1] **8.** -7 [1–1] **9.** -5 [1–1] **10.** -1 [1–2] **11.** 9 [1–2] **12.** 2 [1–2] **13.** 6 [1–2] **14.** -2 [1–2] **15.** 280 [1–3]
16. -120 [1–3] **17.** $-\dfrac{2}{3}$ [1–3] **18.** 4 [1–3] **19.** 180 [1–3] **20.** 30 [1–7] **21.** 18 [1–7] **22.** x^{10} [1–4] **23.** $24x^7$ [1–4]
24. $-30x^3y^6z^2$ [1–4] **25.** $-3x^5y^2$ [1–4] **26.** $\dfrac{4x^3}{y^2}$ [1–4] **27.** $6a$ [1–5] **28.** $-4x$ [1–5] **29.** $2x^2+2x$ [1–5]
30. $3xy^2+7x^2y$ [1–5] **31.** $-20ab+5a+2$ [1–5] **32.** $10x-20y$ [1–6] **33.** $-3x^2-2x+5$ [1–6] **34.** $12x^3-18x^2-24x$ [1–6]
35. $6x^3y^2-15x^2y^4+6xy$ [1–6] **36.** $-6a^3b^2c+10a^2b^2c^3+2ab^2c^2$ [1–6] **37.** $-8x^3+17x^2+10x-3$ [1–7] **38.** $4x^2-12x$ [1–7]
39. $-33x^2+184x$ [1–7] **40.** 75 [1–8] **41.** 8 [1–8] **42.** 45 [1–8] **43.** -3 [1–8] **44.** 88 cm^3 [1–8]

Chapter 1 Practice Test

1. $\dfrac{3}{5}$ [1–1] **2.** -8 [1–1] **3.** -40 [1–2] **4.** $-\dfrac{2}{11}$ [1–2] **5.** 41 [1–3] **6.** 41 [1–3] **7.** -16 [1–3] **8.** -28 [1–4] **9.** $\dfrac{2}{5}$ [1–4]
10. -44 [1–4] **11.** -243 [1–5] **12.** $-24x^6y^7$ [1–5] **13.** $\dfrac{a^2}{3}$ [1–5] **14.** $-\dfrac{4x^4}{y}$ [1–5] **15.** $13x-13y$ [1–6] **16.** a^2b+6ab^2+7ab
[1–6] **17.** $21x-3y+2$ [1–7] **18.** $7x^2-17x$ [1–7] **19.** $20a^3b^2-8a^4b^3-20a^2b^4$ [1–6] **20.** 30 [1–8] **21.** 36 cm [1–8]

Chapter 2 Survey

1. yes [2–1] **2.** $x = -2$ [2–2] **3.** $x = -4$ [2–2] **4.** $x = 12$ [2–2] **5.** $x = \dfrac{9}{7}$ [2–2] **6.** $x = -4$ [2–3] **7.** $x = \dfrac{11}{3}$ [2–3]

8. $x = \dfrac{2a}{17}$ [2–4] **9.** $x \le 10$ [2–5] **10.** $x < -2$ or $x > 8$ [2–6]

Exercise 2-1-1

1. true **3.** conditional **5.** conditional **7.** yes **9.** yes

Exercise 2-1-2

1. equivalent **3.** not equivalent **5.** not equivalent **7.** not equivalent **9.** equivalent

Exercise 2-2-1

1. $x = 2$ **3.** $x = 10$ **5.** $x = 2$ **7.** $x = -14$ **9.** $x = 13$

Exercise 2-2-2

1. $x = 12$ **3.** $x = \dfrac{2}{5}$ **5.** $x = -\dfrac{1}{2}$ **7.** $x = -\dfrac{1}{12}$ **9.** $x = 5$ **11.** $x = 12$ **13.** $x = -6$ **15.** $x = -125$ **17.** $x = 64$ **19.** $x = \dfrac{6}{5}$

Exercise 2-3-1

1. $x = 1$ **3.** $y = -9$ **5.** $a = -4$ **7.** $x = 2$ **9.** $x = \dfrac{10}{3}$ **11.** $y = \dfrac{7}{3}$ **13.** $y = \dfrac{14}{5}$ **15.** $y = 5$ **17.** $x = \dfrac{2}{15}$ **19.** $b = \dfrac{45}{4}$

21. $x = -5$ **23.** $x = \dfrac{46}{45}$ **25.** $x = \dfrac{5}{9}$ **27.** $x = -\dfrac{40}{7}$ **29.** $x = -\dfrac{1}{5}$ **31.** $w = 11$ meters **33.** $d \approx 49$ yards

Exercise 2-4-1

1. $x = 2y$ **3.** $x = 2y$ **5.** $x = -6a$ **7.** $x = -\dfrac{11a}{9}$ **9.** $x = -\dfrac{b}{4}$ **11.** $y = \dfrac{x}{5}$ **13.** $a = -\dfrac{x}{6}$ **15.** $a = 29x + 75$ **17.** $r = \dfrac{I}{pt}$

19. $b = \dfrac{2A - hc}{h}$ **21.** $v = \dfrac{2s - 2h - gt^2}{2t}$

Exercise 2-5-1

1. $6 < 10$ **3.** $-3 < 3$ **5.** $4 > 1$ **7.** $-2 > -3$

9. $0 < 7$

Exercise 2-5-2

1. **3.** **5.** **7.** **9.**

11. $x > 2$ **13.** $x < 0$ **15.** $x \ge 1$ **17.** $x < 3$ **19.** $-5 < x \le 2$

Exercise 2-5-3

1. $x < 4$ **3.** $x < -5$ **5.** $x < -3$ **7.** $x < 4$

9. $x \le 3$ **11.** $x \ge -1$ **13.** $x \ge -2$

15. $x > 3$ **17.** $x < \dfrac{1}{2}$ **19.** $x > \dfrac{45}{13}$

Exercise 2-6-1

1. $x = -3, x = 3$ **3.** $x = -13, x = 3$ **5.** $x = 1, x = 11$ **7.** $x = -3, x = 3$ **9.** $x = -\dfrac{5}{3}, x = 3$

Exercise 2-6-2

1. $-3 < x < 3$ **3.** $-13 < x < 3$ **5.** $-3 \le x \le 7$

7. $-2 < x < 1$ **9.** $x < \dfrac{1}{2}$ or $x > 1$ **11.** $x \le -\dfrac{4}{3}$ or $x \ge 2$

Chapter 2 Review

1. equivalent [2–1]   **2.** not equivalent [2–1]   **3.** not equivalent [2–1]   **4.** $x = -5$ [2–2]   **5.** $x = -10$ [2–2]   **6.** $y = \dfrac{1}{9}$ [2–2]

7. $y = -75$ [2–2]   **8.** $x = \dfrac{9}{8}$ [2–2]   **9.** $x = -3$ [2–3]   **10.** $x = 8$ [2–3]   **11.** $x = 6$ [2–3]   **12.** $q = 4$ [2–3]   **13.** $p = \dfrac{35}{6}$ [2–3]

14. $x = \dfrac{5-a}{10}$ [2–4]   **15.** $x = -\dfrac{8a}{7}$ [2–4]   **16.** $a = -35x + 24$ [2–4]   **17.** $y = \dfrac{P-2x}{2}$ [2–4]   **18.** $z = \dfrac{Hxy - w}{y}$ [2–4]

19. $x = -5$ or $x = 13$ [2–6]   **20.** $x = -\dfrac{5}{3}$ or $x = 1$ [2–6]   **21.** $x \le 5$ [2–5]   **22.** $x \ge -8$

[2–5]   **23.** $x \le -3$ [2–5]   **24.** $x < -6$ [2–5]   **25.** $x \ge -\dfrac{41}{2}$ [2–5]

26. $-\dfrac{3}{2} < x < -1$ [2–6]   **27.** $x \le -1$ or $x \ge 4$ [2–6]   **28.** $-6 \le x \le 9$

[2–6]   **29.** $r = \$7$ per hour [2–3]   **30.** $F = 68°$ [2–3]   **31.** $l = 8\dfrac{3}{4}$ meters [2–3]   **32.** $h \approx 6$ ft [2–3]

Chapter 2 Practice Test

1. d [2–1]   **2.** $x = -12$ [2–2]   **3.** $x = 6$ [2–3]   **4.** $x = -5$ [2–3]   **5.** $x = 9$ [2–3]   **6.** $x = -5$ [2–3]   **7.** $x = 12$ [2–3]

8. $x = \dfrac{6-5y}{3}$ [2–4]   **9.** $x = \dfrac{5a}{9}$ [2–4]   **10.** $x = -\dfrac{11}{3}, x = 3$ [2–6]   **11.** [2–5]   **12.** [2–5]

13. [2–5]   **14.** $x > 0$ [2–5]   **15.** $x \ge \dfrac{29}{2}$ [2–5]   **16.** $-\dfrac{8}{5} \le x \le 6$ [2–6]   **17.** $x > 3$ or $x < -\dfrac{13}{3}$ [2–6]

18. $w = 5$ cm [2–3]   **19.** $b = \dfrac{2A - ha}{h}$ [2–4]   **20.** $x \le \dfrac{14}{3}$ [2–5]

Chapter 3 Survey

1. a. $2w - 9$ **b.** $\dfrac{1}{3}x + 1$ **c.** $0.173p$ **d.** $10d$ [3–1]   **2.** 6 yds, 12 yds, 8 yds [3–2]   **3.** 10 students [3–2]   **4.** 4 hours [3–3]   **5.** 25 miles [3–2]   **6.** 2 liters [3–4]

Exercise 3-1-1

1. $n + 20$ **3.** $2x$ **5.** $t - 4$ **7.** $\dfrac{1}{2}x$ **9.** $2h - 3$ **11.** $0.06x$ **13.** $25q + 20$ **15.** $4x$

Exercise 3-1-2

1. Let $z =$ Zack's age; $z + 6 =$ Becky's age **3.** Let $n =$ the number of students in the 11:00 A.M. class; $n - 8 =$ the number of students in the 9:00 A.M. class **5.** Let $w =$ the width of the garden; $2w + 5 =$ the length of the garden **7.** Let $x =$ the first investment; $10,000 - x =$ the second investment **9.** Let $x =$ this year's enrollment; $0.93x =$ last year's enrollment

11.

Unknown	Expression
Amount of flour in brownies	$3x$
Amount of flour in pancakes	$6x$
Amount of flour in crepes	x

13.

Unknown	Expression
# of dollars Anita has	x
# of dollars Ryan has	$x - 5$
# of dollars Lucas has	$x + 13$

15. Let $x =$ the original amount; $0.05x =$ the interest earned; $1.05x =$ the total amount at the end of the year

Exercise 3-2-1

1. 7 inches and 15 inches **3.** 9:00 class: 38 students; 11:00 class: 46 students **5.** Width: 6 m; Length: 17 m **7.** $3372 and $6628 **9.** Lemieux scored 690 goals; Gretzky scored 894 goals. **11.** Retail price: $26.80; Tax: $1.34 **13.** Stocks: $39,250; Bonds: $47,250 **15.** 11% investment: $5500; 13% investment: $4500 **17.** 20 teeth **19.** Height: 8 cm **21.** Nicholas' weight: 70 kg; Matthew's weight: 75 kg; Adam's weight: 77 kg **23.** 8 ounces

Exercise 3-3-1

1. 264 miles **3.** 19 mph **5.** 4:00 PM **7.** 2 hours **9.** $\dfrac{1}{2}$ hour **11.** Chris: 6 mph; Beth: 15 mph **13.** Nancy: 55 mph; Barb: 65 mph; 137.5 miles **15.** 390 km/hr **17.** 40 km/hr

Exercise 3-4-1

1. 10 lbs of caramels; 20 lbs of creams **3.** 18 nickels; 57 quarters **5.** 63 nickels; 21 quarters **7.** 100 gallons of skim milk; 125 gallons of 3.6% milk **9.** 60 bags of fertilizer; 40 bags of weed killer **11.** 20 ml **13.** 10 oz of Keemun; 20 oz of Oolong; 20 oz of Assam **15.** 2 liters

Chapter 3 Review

1. $x + 5$ [3–1] **2.** $25x$ [3–1] **3.** $\dfrac{y}{100}$ [3–1] **4.** x; $84 - x$ [3–1] **5.** w; $3w + 7$ [3–1] **6.** $0.08A$ [3–1] **7.** $1.102x$ [3–1]

8. 24 cm; 12 cm; 28 cm [3–2] **9.** 47 cm; 53 cm [3–2] **10.** Carolyn's age: 28; Mindy's age: 35 [3–2] **11.** $48 [3–2]

12. $1500 [3–2] **13.** Julia's age: 14; Jordan's age: 18; Shasta's age: 26 [3–2] **14.** 4 nickels; 8 dimes; 10 quarters [3–2]

15. There are 25 algebra students, 50 psychology students, and 47 history students. [3–2] **16.** 126 km [3–3] **17.** 40 grams [3–4]

18. 80 km/hr [3–3] **19.** 4.5 m; 13.5 m; 22 m [3–2] **20.** 120 km/hr [3–3] **21.** 6 ft by 16 ft [3–2] **22.** $\dfrac{2}{3}$ liter [3–4]

23. $\dfrac{1}{2}$ hour [3–3] **24.** $\dfrac{1}{4}$ liter [3–4] **25.** 16 oz [3–2] **26.** 18 oz; 24 oz [3–2]

Chapter 3 Practice Test

1. a. $3w + 6$ **b.** $\frac{1}{2}x - 7$ **c.** $0.094x$ **d.** $\frac{d}{7}$ [3–1] **2.** 5 nickels; 8 dimes; 16 quarters [3–4] **3.** 12 m; 24 m; 27 m [3–2]

4. 3 km/hr [3–3] **5.** $1\frac{3}{5}$ oz [3–2] **6.** 5 liters [3–4]

Chapter 4 Survey

1. $2x^2 + 9x - 5$ [4–1] **2.** $x^3 + x^2 - 11x + 10$ [4–1] **3.** $7x(5x - 4)$ [4–2] **4.** $2x^2 y^2 \left(3xy - 4x + 5y\right)$ [4–2] **5.** $(x - y)(a - 1)$ [4–2]

6. $(x + 3)(a - b)$ [4–3] **7.** $(x - 8)(x - 5)$ [4–4] **8.** $(3x + 5)(x + 4)$ [4–4] **9.** $(2x + 5)(2x - 5)$ [4–6] **10.** $\left(x + 10\right)^2$ [4–6]

11. $3(2x + 1)(x - 5)$ [4–7] **12.** $4x\left(2x - 3\right)^2$ [4–7] **13.** $\left(2a - 1\right)\left(4a^2 + 2a + 1\right)$ [4–6] **14.** no [4–8] **15.** yes [4–8]

Exercise 4-1-1

1. $x^3 + 8x^2 + 16x + 5$ **3.** $x^3 - 125$ **5.** $2x^3 - 13x^2 + 7x - 6$ **7.** $15a^3 + 8a^2 b - 19ab^2 - 4b^3$

9. $x^3 - x^3 y + 5x^2 + x^2 y - x^2 y^2 + 4xy + x + 5$

Exercise 4-1-2

1. $a^2 + 5a + 6$ **3.** $x^2 + 7x + 10$ **5.** $a^2 - 5a + 6$ **7.** $a^2 - 8a + 15$ **9.** $x^2 + 4x - 5$ **11.** $a^2 + 4a - 45$ **13.** $6x^2 + 7x + 2$

15. $6a^2 + 5a - 4$ **17.** $6x^2 - 19x + 15$ **19.** $10a^2 + 6a - 5ab - 3b$ **21.** $4a^2 + ac - 20a - 5c$ **23.** $m^2 + 10m + 25$

Exercise 4-1-3

1. $x^2 - 9$ **3.** $4x^2 - 49$ **5.** $x^2 - y^2$ **7.** $x^2 + 2xy + y^2$ **9.** $x^2 - 12x + 36$ **11.** $25x^2 - 20x + 4$ **13.** $x^3 + 27$ **15.** $6a^2 + a - 35$

17. $4y^2 - 9$ **19.** $4x^2 - 4x + 1$

Exercise 4-2-1

1. $2(6a + 5b)$ **3.** $a(a + 5)$ **5.** $2a(a + 3)$ **7.** $x(x - 1)$ **9.** $4xy(2x + 3)$ **11.** $2ab(5a - b + 3)$ **13.** prime **15.** $7a(2ab - 5b - 9)$

17. $11ab^3\left(a^2 b + 4 - 3ab\right)$ **19.** prime

Exercise 4-2-2

1. $(x + 4)(2x + 3)$ **3.** $(a - 1)(a + 5)$ **5.** $(y + 4)(6x - 7)$ **7.** $(x - 2)(3x + 1)$ **9.** $(a - 4)(7a - 15)$ **11.** $(x + 3)(x - 1)$

13. $x(x + 1)(3x + 2)$ **15.** $a(a + b)(4 + a)$ **17.** $(x - 1)^2$ **19.** $2a(a + 4)(4a + 1)$

Exercise 4-3-1

1. $(x + y)(a + 3)$ **3.** $(b + c)(a - 2)$ **5.** $(2x + y)(a + 3)$ **7.** $(x - y)(a + 2)$ **9.** $(x + 2y)(5 + a)$ **11.** $(x - y)(3a + 2)$

13. $(x + y)(2a - 1)$ **15.** $(2a - 1)(3x + y)$ **17.** $(x - 2)(y + 3)$ **19.** $(y - 2)(x + 5)$

Exercise 4-4-1

1. $(x + 2)(x + 3)$ **3.** $(x + 3)(x - 2)$ **5.** $(x - 6)(x + 1)$ **7.** $(x + 9)(x - 2)$ **9.** $(y - 6)(y - 2)$ **11.** $(p + 35)(p - 1)$ **13.** prime

15. $(a - 5b)(a - 7b)$ **17.** $(x + 16)(x - 3)$ **19.** $(x - 8)(x - 5)$ **21.** $(x + 10)(x + 5)$ **23.** $(x + 50)(x - 1)$ **25.** $(t + 4)(t + 18)$

27. $(x-8y)(x+3y)$ **29.** $(x-4)(x-24)$ **31.** $(x+28)(x+3)$ **33.** $(x-84)(x+1)$ **35.** $(x-35)(x-4)$ **37.** $(x^3-5)(x^3+4)$

39. $(y^4+6)(y^4-5)$

Exercise 4-4-2

1. $(2x+1)(x+2)$ **3.** $(2x+3)(x+1)$ **5.** $(2y+3)(5y+2)$ **7.** $(3c+2)(2c+1)$ **9.** $(4x-3)(x-2)$ **11.** $(8x+3)(4x-1)$ **13.** prime

15. $(6x+1)(x+5)$ **17.** $(5a+3)(a+1)$ **19.** $(8x+y)(2x-y)$ **21.** $(4a-9b)(a+2b)$ **23.** $(5x^2-12)(3x^2+2)$

Exercise 4-5-1

1. $(x+4)(x+5)$ **3.** $(x-2)(x-6)$ **5.** $(3x+2)(x+1)$ **7.** $(5x-1)(x+2)$ **9.** $(3x-2)(x-4)$ **11.** $(4x-3)(x+3)$

13. $(2x+1)(2x+3)$ **15.** $(4x+1)(4x-3)$

Exercise 4-6-1

1. $(x+2)(x-2)$ **3.** $(a+9)(a-9)$ **5.** $(y+11)(y-11)$ **7.** $(4x+1)(4x-1)$ **9.** $(2x+5y)(2x-5y)$ **11.** $(2+x)(2-x)$

13. $(4a+11b)(4a-11b)$ **15.** $(x^2+3)(x^2-3)$ **17.** $(5a^2+7)(5a^2-7)$

Exercise 4-6-2

1. $(x+y)(x^2-xy+y^2)$ **3.** $(a-2)(a^2+2a+4)$ **5.** $(x-1)(x^2+x+1)$ **7.** $(3a+1)(9a^2-3a+1)$ **9.** $(3a+2b)(9a^2-6ab+4b^2)$

11. $(4x-3y)(16x^2+12xy+9y^2)$

Exercise 4-6-3

1. $(x+3)^2$ **3.** $(x-7)^2$ **5.** $(2x+3)^2$ **7.** $(3x+5)^2$ **9.** $(6x-5y)^2$

Exercise 4-7-1

1. $2(2x+1)(x+3)$ **3.** $3(x+8)(x-6)$ **5.** $3(x+2)(x-2)$ **7.** prime **9.** $7(3b+1)(2b-1)$ **11.** $x(x-2)(x-6)$

13. $(a+3)(x+1)(x-1)$ **15.** $2(2x-1)^2$ **17.** $(m+7)(m-2)$ **19.** $(2x-3)(2x+1)$ **21.** $3x(3x+5)(3x-5)$

23. $(x+6)(x-5)$ **25.** prime **27.** $5(x-1)(x^2+x+1)$

Exercise 4-8-1

1. yes **3.** no **5.** no **7.** no **9.** $(2x+1)(2x-1)(x+2)$

Chapter 4 Review

1. $2x^2-x-6$ [4–1] **2.** $9a^2-12a+4$ [4–1] **3.** $9x^2-64$ [4–1] **4.** $2x^3+5x^2+x+12$ [4–1] **5.** x^2+2x-y^2-4y-3 [4–1]

6. $3(3x+2y)$ [4–2] **7.** $a(a-1)$ [4–2] **8.** $2x^2y(3x+5y)$ [4–2] **9.** $3ab(2a+3b-2)$ [4–2] **10.** $(a+b)(x-2)$ [4–2]

11. $(5x+7)(5x-7)$ [4–6] **12.** $(x+4)^2$ [4–6] **13.** $(3x-5)^2$ [4–6] **14.** $5(3+x)(3-x)$ [4–7] **15.** $3(2x-1)(4x^2+2x+1)$ [4–7]

16. $(x+2)(x+3)$ [4–4] **17.** $(x+7)(x-2)$ [4–4] **18.** $(x-1)(x-3)$ [4–4] **19.** $(5+x)(3-x)$ [4–7] **20.** $(9+x)(8-x)$ [4–7]

21. prime [4–4] **22.** $(2x-3)(x+1)$ [4–4] **23.** $(3a-2)(a+5)$ [4–4] **24.** $(3x-4)(2x+1)$ [4–4] **25.** $(6t-1)(t-2)$ [4–4]

26. $2(2m+1)(m+3)$ [4–7] **27.** $3(3x+2)(2x-1)$ [4–7] **28.** $2x(x+2)(3x-2)$ [4–7] **29.** $3(x^4+3x^2+5)$ [4–2]

30. $\left(5z^3+3\right)\left(z^3+2\right)$ [4–4] **31.** $(x+y)(c+d)$ [4–3] **32.** $(x+y)(a-3)$ [4–3] **33.** $(x-5)(a-4)$ [4–3] **34.** $(a+2)(a-2)(x+2)$ [4–7] **35.** $(a+1)(a-1)(2x+3)$ [4–7] **36.** yes [4–8] **37.** yes [4–8] **38.** no [4–8] **39.** yes [4–8] **40.** no [4–8]

Chapter 4 Practice Test

1. $6x^2+7x-3$ [4–1] **2.** $2x^3-5x^2+8x-3$ [4–1] **3.** $11a(2a+3)$ [4–2] **4.** $2xy\left(5x^2-3xy+y^2\right)$ [4–2] **5.** $(a-1)(2x-3)$ [4–2]

6. $(p+9)(p-9)$ [4–6] **7.** $(x-11)^2$ [4–6] **8.** $(x+3)(a+b)$ [4–3] **9.** $(x-7)^2$ [4–6] **10.** $(4a-3)(a-4)$ [4–4]

11. $(2x+3)(x+5)$ [4–4] **12.** $(m+9)(m+2)$ [4–4] **13.** $(5x+12)(5x-12)$ [4–6] **14.** $(r-9)(r+3)$ [4–4]

15. $(a+3)(x+3)(x-3)$ [4–7] **16.** $2(3x+1)(x-2)$ [4–7] **17.** $8x(2x+3)(2x-3)$ [4–7] **18.** $(z+9)(z-7)$ [4–4]

19. $3y(2y-3)(y+2)$ [4–7] **20.** $3x(3x+2)(3x+2)$ [4–7] **21.** $(2t+3)(3t-5)$ [4–4] **22.** $4(3x+5)(2x+1)$ [4–7] **23.** yes [4–8]

24. no [4–8]

Mid Book Test

1. 2 [1–3] **2.** $8x^6$ [1–4] **3.** $15x^3y^4-10x^4y$ [1–6] **4.** -27 [1–7] **5.** $20x^2-3x$ [1–7] **6.** 9 [1–8] **7.** -6 [2–3] **8.** 12 [2–3]

9. 5 [2–3] **10.** $x=\dfrac{9}{5}a$ [2–4] **11.** $x<-4$ [2–5] **12.** $-1\le x\le 9$ [2–6]

13. $7a^2b^2\left(2b-a^2+3ab\right)$ [4–2] **14.** $2(3y+2)(3y-2)$ [4–7] **15.** $\left(2m-5\right)^2$ [4–6] **16.** $7(2x-1)(x+2)$ [4–7]

17. $(a-b)(x+1)(x-1)$ [4–7] **18.** $\left(3p+1\right)\left(9p^2-3p+1\right)$ [4–6] **19.** yes;

$$\require{enclose}\begin{array}{r}x^2+\ 5x-2\\ x-3\enclose{longdiv}{x^3+2x^2-17x+6}\\ \underline{x^3-3x^2}\\ 5x^2-17x\\ \underline{5x^2-15x}\\ -2x+6\\ \underline{-2x+6}\\ 0\end{array}$$

[4–8] **20.** $\dfrac{m}{60}$ [3–1]

21. Stocks: \$3500 Bonds: \$7000 [3–2] **22.** 61 cm; 49 cm; 41 cm [3–2]

23. 35 mph [3–3] **24.** 10 litres [3–4]

Chapter 5 Survey

1. $\dfrac{x+3}{x+2}$ [5–1] **2.** $\dfrac{2x-1}{x+3}$ [5–2] **3.** $\dfrac{y}{y-3}$ [5–2] **4.** $(x+4)(x+6)(x-4)$ [5–3] **5.** $\dfrac{2x^2-13x+21}{(x+3)(2x-7)}$ [5–3]

6. $\dfrac{x^2+12x+7}{(x+4)(x-2)(x-1)}$ [5–4] **7.** $\dfrac{6}{(x+3)(x-1)}$ [5–5] **8.** $\dfrac{1}{x(x-3)}$ [5–6] **9.** $x=2$ [5–7] **10.** 2 hours [5–7]

Exercise 5-1-1

1. $\dfrac{x}{x+1}$ **3.** $\dfrac{x+1}{x+2}$ **5.** $\dfrac{y+1}{y-5}$ **7.** $\dfrac{x-1}{x-3}$ **9.** in simplest form **11.** $\dfrac{x}{x+3}$ **13.** $\dfrac{a+5}{a-3}$ **15.** $-\dfrac{x+3}{x+4}$ **17.** $\dfrac{x-3}{x+2}$ **19.** $\dfrac{3a+2b}{2a-3b}$

21. $\dfrac{9x^2-3x+1}{x-1}$

Exercise 5-2-1

1. $\dfrac{x+1}{x-3}$ **3.** $\dfrac{1}{a+1}$ **5.** $\dfrac{1}{x-1}$ **7.** $(m+1)(m+3)^2$ **9.** $\dfrac{x-1}{(x-2)(x+1)}$ **11.** $\dfrac{12x+5}{x+7}$ **13.** $\dfrac{2x}{(x-1)^2}$ **15.** $t+4$ **17.** $\dfrac{1}{p+1}$ **19.** $\dfrac{1}{x-9}$

21. $\dfrac{2(x+1)}{x+2}$ **23.** $\dfrac{3x+5}{2x-1}$ **25.** 1 **27.** $\dfrac{(b+2)(b^2-2b+4)}{b(b-2)}$ **29.** $\dfrac{x+6}{(x+2)(x+4)}$

Exercise 5-3-1

1. $\dfrac{5}{7}$ **3.** $\dfrac{8}{x}$ **5.** $\dfrac{1-a}{y}$ **7.** $\dfrac{a-1}{x}$ **9.** $\dfrac{a+b+c}{x}$ **11.** $\dfrac{4-x}{x}$

Exercise 5-3-2

1. xyz **3.** $18a^2b^2$ **5.** $x(x+2)$ **7.** $a^2(a-2)$ **9.** $(a+3)(a-3)$ **11.** $(b-4)^3(b+3)^3$ **13.** $(2x+3)(x+4)(2x-3)$

15. $(y-3)^2(y+2)(y-4)$

Exercise 5-3-3

1. $\dfrac{ab^2}{a^2b^2}$ **3.** $\dfrac{x-3}{(x+2)(x-3)}$ **5.** $\dfrac{x+2}{x(x+2)}$ **7.** $\dfrac{2a^2-13a+6}{(a-6)(3a+4)}$ **9.** $\dfrac{x^2-3x-10}{(x+5)(x-3)(x+2)}$

Exercise 5-4-1

1. $\dfrac{y+x}{xy}$ **3.** $\dfrac{x+11}{3(x+5)}$ **5.** $\dfrac{y^2+y+3}{y(y+3)}$ **7.** $\dfrac{c+6}{6(c-3)}$ **9.** $\dfrac{b-1}{(b+2)(b-3)}$ **11.** $\dfrac{2x+1}{x(x+2)}$ **13.** $\dfrac{5x+3}{(x+3)^2(x-3)}$

15. $\dfrac{3b^2-7}{(b+2)(b+1)(b-3)}$ **17.** $\dfrac{6}{x+1}$ **19.** $\dfrac{2p^2-4p-2}{(p+2)(p-1)}$ **21.** $\dfrac{p^2-11p+4}{(p+4)^2(p-4)}$ **23.** $\dfrac{4}{x-2}$

Exercise 5-5-1

1. $\dfrac{3x-20}{5x}$ **3.** $\dfrac{x-2}{3(x+1)}$ **5.** $\dfrac{12-2x}{x(x+3)}$ **7.** $\dfrac{2b^2+6b+12}{(b+5)(b-3)}$ **9.** $\dfrac{p^2-p-1}{p(p+2)}$ **11.** $\dfrac{5x^2-13x+11}{(x+3)(x-2)}$ **13.** $\dfrac{-3y^2+16y-17}{3(y+2)(y-2)}$

15. $\dfrac{4}{(x+1)(x+5)}$ **17.** $\dfrac{y^2+13y-21}{(y+1)(y-5)(y-2)}$ **19.** $-\dfrac{1}{5}$

Exercise 5-6-1

1. a **3.** $\dfrac{1}{xy}$ **5.** $\dfrac{s-r}{r^2s^2}$ **7.** $\dfrac{2}{xy}$ **9.** $\dfrac{2d+3c}{d-c}$ **11.** $\dfrac{a+1}{a^2b}$ **13.** $\dfrac{2x-1}{x^2+x}$ **15.** $\dfrac{a^3+a^2b-ab^2-b^3}{ab}$ **17.** $\dfrac{-1}{p(p+4)}$ **19.** $\dfrac{x^2}{-2}$

Exercise 5-7-1

1. $x=10$ **3.** $a=-\dfrac{9}{8}$ **5.** $c=4$ **7.** $b=-2$ **9.** no solution **11.** $x=3$ **13.** $p=0$ **15.** $b=-4$ **17.** $x=-1$ **19.** $3\dfrac{3}{4}$ hours

(3 hours and 45 minutes) **21.** $1\dfrac{1}{2}$ hours **23.** $8\dfrac{4}{7}$ hours **25.** 8 days

Chapter 5 Review

1. $\dfrac{x+15}{4}$ [5–1] **2.** $\dfrac{2}{3}$ [5–1] **3.** $\dfrac{y+3}{y+8}$ [5–1] **4.** $\dfrac{a-7}{a+7}$ [5–1] **5.** $\dfrac{3}{5(x-1)}$ [5–1] **6.** $x(x+2)$ [5–3] **7.** $x(x+3)$ [5–3]

8. $y(y-5)(y+1)$ [5–3] **9.** $(a+1)(a-2)(a+2)$ [5–3] **10.** $(x+1)(x-1)(x-2)$ [5–3] **11.** $\dfrac{5}{7}$ [5–2] **12.** $\dfrac{t-1}{t-4}$ [5–2] **13.** $\dfrac{1-x}{x}$

[5–2] **14.** $\dfrac{2x-5}{x+4}$ [5–2] **15.** $\dfrac{2}{3}$ [5–2] **16.** $\dfrac{-1}{(x+1)(x-1)}$ [5–2] **17.** $\dfrac{x+2}{x+4}$ [5–2] **18.** $\dfrac{2x+3}{2x+5}$ [5–2] **19.** $\dfrac{5x+6}{x(x+2)}$ [5–4]

20. $\dfrac{t^2+4}{t(t+3)}$ [5–4]   **21.** $\dfrac{x^2+2x+3}{x(x-5)(x+1)}$ [5–4]   **22.** $\dfrac{9b^2-4b-25}{(b+1)(b+2)(b-2)}$ [5–4]   **23.** $\dfrac{2x-3}{(x-1)(x-2)}$ [5–4]   **24.** $\dfrac{15-x}{4(x+5)}$ [5–5]

25. $\dfrac{s-8}{(s+1)(s-2)}$ [5–5]   **26.** $\dfrac{3}{b-1}$ [5–5]   **27.** $\dfrac{x^2+6x-8}{(x+2)(x-2)(x+6)}$ [5–5]   **28.** $\dfrac{-17p-25}{(p+1)(p+2)(p-7)}$ [5–5]   **29.** $\dfrac{1}{ab}$ [5–6]

30. $\dfrac{1}{a}$ [5–6]   **31.** $\dfrac{1-x-2x^2}{x}$ [5–6]   **32.** $\dfrac{x+1}{x(x+2)}$ [5–6]   **33.** $\dfrac{x^2+2xy+y^2}{xy}$ [5–6]   **34.** $x=1$ [5–7]   **35.** $a=\dfrac{7}{6}$ [5–7]

36. no solution [5–7]   **37.** $x=\dfrac{1}{3}$ [5–7]   **38.** $x=-1$ [5–7]   **39.** 20 ohms [5–7]   **40.** 6 hours 40 minutes [5–7]

Chapter 5 Practice Test

1. $\dfrac{x+3}{4}$ [5–1]   **2.** $\dfrac{a+3}{a+5}$ [5–1]   **3.** $\dfrac{x+5}{x-1}$ [5–2]   **4.** $\dfrac{2}{3}$ [5–2]   **5.** $x(x-7)$ [5–3]   **6.** $(x+6)(x-6)(x+4)$ [5–3]   **7.** $\dfrac{5x+2}{x(x+1)}$ [5–4]

8. $\dfrac{3z^2+11z-3}{(z+4)(z+2)}$ [5–5]   **9.** $\dfrac{1}{a}$ [5–6]   **10.** $\dfrac{-3}{y(y+2)}$ [5–6]   **11.** $x=2$ [5–7]   **12.** 1 minute 20 seconds [5–7]

Chapter 6 Survey

1. x^4y^8 [6–1]   **2.** 2 [6–1]   **3.** $\dfrac{x^2}{y^6}$ [6–1]   **4.** $\dfrac{-4}{125x^{10}}$ [6–2]   **5.** 3.5×10^6 [6–2]   **6.** -4 [6–3]   **7.** 9 [6–4]   **8.** x^3y^2 [6–5]   **9.** $x^{\frac{7}{6}}$

[6–4]   **10.** $-10\sqrt{3}$ [6–5]   **11.** $3\sqrt{10}+\sqrt{15}$ [6–6]   **12.** -163 [6–6]   **13.** $\dfrac{3\sqrt[3]{x^2}}{x}$ [6–7]   **14.** $\dfrac{4\sqrt{5}-4\sqrt{2}}{3}$ [6–7]

Exercise 6-1-1

1. x^8 **3.** x^{10} **5.** x^4y^3 **7.** x^3 **9.** $\dfrac{1}{x}$

Exercise 6-1-2

1. x^{10} **3.** x^8y^4 **5.** $\dfrac{m^6}{n^9}$ **7.** $16a^8$ **9.** $\dfrac{8}{x^6y^3}$ **11.** $\dfrac{y^2}{x^2}$ **13.** $8x^6y^6$ **15.** $64a^{18}b^6$ **17.** $-3x^2$ **19.** $\dfrac{-y^3}{5x}$ **21.** $-2s^8t^4$ **23.** $-4x^8y^9$

25. $288x^{22}y^{16}$ **27.** $-\dfrac{2}{x}$ **29.** $-\dfrac{x^4}{5y^4}$

Exercise 6-2-1

1. 1   **3.** 1   **5.** $\dfrac{1}{a^5}$ **7.** $\dfrac{1}{27}$ **9.** $\dfrac{1}{8x^3}$ **11.** 36   **13.** a^5 **15.** $\dfrac{16}{9}$ **17.** $\dfrac{x^6}{4}$

Exercise 6-2-2

1. $\dfrac{1}{x^4}$ **3.** $\dfrac{1}{y^6}$ **5.** 81   **7.** y^7 **9.** $\dfrac{1}{32}$ **11.** $-\dfrac{4}{z}$ **13.** x^8 **15.** $\dfrac{1}{x^{15}}$ **17.** $\dfrac{1}{8x^6}$ **19.** $\dfrac{1}{8}$ **21.** $\dfrac{1}{y^7}$ **23.** $\dfrac{z^4}{y^4}$

Exercise 6-2-3

1. $\dfrac{-9}{8x^{11}}$ **3.** $2x^2y$ **5.** $-512x^{20}$ **7.** $\dfrac{1}{ab}$ **9.** $\dfrac{b+a}{ab}$ **11.** $\dfrac{4x}{4+x}$

Exercise 6-2-4

1. yes **3.** yes **5.** yes **7.** no **9.** yes **11.** 5×10^3 **13.** 2.35×10^{-7} **15.** 5.2×10^{-9} **17.** 6.8×10^1 **19.** 7.28×10^5 **21.** $320,100$

23. 0.00000728 **25.** $50,200,000,000$ **27.** 0.0000407 **29.** 3.6 **31.** 4.9×10^7 miles **33.** $60,000,000,000,000$ miles

35. 0.0000000000035 cm **37.** 4.3×10^{19}

Exercise 6-3-1

1. 2 **3.** 3 **5.** -1 **7.** $\dfrac{5}{4}$ **9.** $\dfrac{3}{5}$ **11.** -2 **13.** 3 **15.** -5 **17.** $-\dfrac{6}{7}$ **19.** -9 **21.** -6 **23.** -1

Exercise 6-3-2

1. $2 < \sqrt{8} < 3; \sqrt{8} \approx 2.828$ **3.** $2 < \sqrt[3]{8.5} < 3; \sqrt[3]{8.5} \approx 2.041$ **5.** $4 < \sqrt{21} < 5; \sqrt{21} \approx 4.583$ **7.** $7 < \sqrt{50} < 8; \sqrt{50} \approx 7.071$

9. $10 < \sqrt{110} < 11; \sqrt{110} \approx 10.488$ **11.** $30 < \sqrt{905} < 31; \sqrt{905} \approx 30.083$

Exercise 6-4-1

1. $\sqrt[3]{x}$ **3.** $-\sqrt[4]{a^3}$ **5.** $\sqrt{6}$ **7.** $\dfrac{1}{\sqrt[4]{a^3}}$ **9.** $-\dfrac{1}{\sqrt{3}}$ **11.** $\sqrt[3]{4x^2}$ **13.** $x^{\frac{1}{2}}$ **15.** $x^{\frac{4}{5}}$ **17.** $3^{\frac{1}{5}}$ **19.** $-x^{-\frac{1}{2}}$ **21.** $4^{-\frac{1}{3}}$ **23.** $(2y)^{\frac{1}{3}}$

Exercise 6-4-2

1. 2 **3.** 3 **5.** -9 **7.** 4 **9.** 4 **11.** -4 **13.** 1 **15.** -21

Exercise 6-4-3

1. $x^{\frac{3}{5}}$ **3.** $x^{\frac{1}{12}}$ **5.** $2^{\frac{13}{15}}$ **7.** $10^{\frac{5}{12}}$ **9.** $x^{\frac{11}{12}}$ **11.** $a^{\frac{16}{15}}$ **13.** $72x^6$ **15.** $\dfrac{2}{x^{\frac{5}{2}}}$ **17.** $\sqrt[6]{5}$ **19.** $\sqrt[12]{x^5}$ **21.** $\sqrt[12]{a^{11}}$

Exercise 6-5-1

1. $2\sqrt{2}$ **3.** $3\sqrt{5}$ **5.** $6\sqrt{5}$ **7.** $2\sqrt[3]{2}$ **9.** $-2\sqrt[3]{2}$ **11.** $4\sqrt{5}$ **13.** $21\sqrt{2}$ **15.** $14\sqrt[3]{2}$ **17.** x^2y^3 **19.** $x^2y^2\sqrt{y}$ **21.** $2x^2\sqrt[4]{x}$

23. $2y^2\sqrt[4]{y}$

Exercise 6-6-1

1. $3\sqrt{3}$ **3.** $7\sqrt{5} - 3\sqrt{3}$ **5.** $-3\sqrt[3]{2}$ **7.** 0 **9.** $-\sqrt{6} - \sqrt[3]{2}$ **11.** $2\sqrt{3}$

Exercise 6-6-2

1. $\sqrt{10}$ **3.** $6\sqrt{14}$ **5.** $8a^2$ **7.** $x\sqrt[4]{x}$ **9.** $-40x^2$ **11.** $\sqrt[4]{a^3}$

Exercise 6-6-3

1. $\sqrt{6} + \sqrt{10}$ **3.** $27 + 6\sqrt{15}$ **5.** $30\sqrt{2} + 15\sqrt{5}$ **7.** $12\sqrt{2} - 12 + 12\sqrt{6}$ **9.** $\sqrt{21} + \sqrt{10}$ **11.** $4\sqrt[3]{6} - 12$

Exercise 6-6-4

1. $\sqrt{6} + 2 + \sqrt{15} + \sqrt{10}$ **3.** $6\sqrt{15} - 2\sqrt{6} - 15 + \sqrt{10}$ **5.** -22 **7.** $50 - 4\sqrt{66}$ **9.** $47 + 6\sqrt{10}$ **11.** $187 - 20\sqrt{21}$

Explanation

Exercise 6-7-1

1. $\dfrac{2}{5}$ **3.** $\sqrt{6}$ **5.** 5   **7.** $-\dfrac{\sqrt{5}}{5}$ **9.** $\dfrac{3\sqrt{5}}{5}$ **11.** $\dfrac{\sqrt[3]{2}}{2}$ **13.** $\dfrac{\sqrt{15}}{5}$ **15.** $\sqrt[3]{4}$ **17.** $-\dfrac{\sqrt[4]{y^3}}{y}$ **19.** $2\sqrt[5]{x^2}$ **21.** $\dfrac{\sqrt{x-3}}{x-3}$ **23.** $\dfrac{2\sqrt{2x-4}}{x-2}$

Exercise 6-7-2

1. $\sqrt{2}-1$ **3.** $\dfrac{\sqrt{5}-1}{2}$ **5.** $\sqrt{3}+\sqrt{2}$ **7.** $\dfrac{\sqrt{3}+\sqrt{5}}{-2}$ **9.** $\dfrac{6\sqrt{x}-6y}{x-y^2}$ **11.** $\sqrt{7}+\sqrt{2}$

Exercise 6-7-3

1. $\sqrt[3]{x^2}$ **3.** x^2y^3 **5.** $\sqrt[4]{a^3b}$

Chapter 6 Review

1. $32x^{10}y^{15}$ [6–1]   **2.** $\dfrac{8a^9}{27b^3}$ [6–1]   **3.** $-108x^{14}y^{10}$ [6–1]   **4.** t^6 [6–2]   **5.** 125 [6–2]   **6.** x^3+x [6–2]   **7.** $-\dfrac{1}{8x^6}$ [6–2]   **8.** $-\dfrac{x^7}{4y^4}$ [6–2]   **9.** $-\dfrac{1}{ab}$ [6–2]   **10.** $y^{\frac{1}{6}}$ [6–4]   **11.** $6x^2$ [6–4]   **12.** $\sqrt[6]{a}$ [6–4]   **13.** 5.42×10^7 [6–2]   **14.** 0.000032 [6–2]   **15.** $-\sqrt[3]{a^2}$ [6–4]   **16.** $y^{\frac{1}{5}}$ [6–4]   **17.** 4 [6–3]   **18.** -5 [6–3]   **19.** 2 [6–3]   **20.** 4 [6–3]   **21.** 2 [6–7]   **22.** $\dfrac{3}{2}$ [6–4]   **23.** 9 [6–4]   **24.** $-\dfrac{1}{16}$ [6–4]   **25.** $5\sqrt{2}$ [6–5]   **26.** $-6\sqrt[3]{4}$ [6–5]   **27.** $5a^2\sqrt{3}$ [6–5]   **28.** $-2xy^2\sqrt[3]{5y}$ [6–5]   **29.** $9\sqrt{7}$ [6–6]   **30.** $6\sqrt{6}-7\sqrt{3}$ [6–6]   **31.** -150 [6–6]   **32.** $60x^4\sqrt{3}$ [6–6]   **33.** $30-2\sqrt{15}$ [6–6]   **34.** $26-8\sqrt{3}$ [6–6]   **35.** $\dfrac{\sqrt{2}}{3}$ [6–7]   **36.** $\dfrac{3\sqrt[3]{2}}{2}$ [6–7]   **37.** $\sqrt{a-b}$ [6–7]   **38.** $\dfrac{-\sqrt{6}-\sqrt{2}}{2}$ [6–7]   **39.** $\sqrt[4]{a^3}$ [6–7]

Chapter 6 Practice Test

1. -6 [6–7]   **2.** 25 [6–5]   **3.** 4 [6–7]   **4.** $\dfrac{5}{7}$ [6–7]   **5.** 5.61×10^{-6} [6–6]   **6.** 380,000,000 [6–6]   **7.** $x^{24}y^{36}$ [6–3]   **8.** $\dfrac{1}{4x^7y}$ [6–4]   **9.** x^4 [6–4]   **10.** x^6+x^2 [6–4]   **11.** $\dfrac{1}{a^{16}}$ [6–2]   **12.** $-10m^{\frac{13}{12}}$ [6–4]   **13.** $7\sqrt{2}$ [6–2]   **14.** $20\sqrt{2}$ [6–2]   **15.** $28-\sqrt{14}$ [6–2]   **16.** $8+2\sqrt{15}$ [6–6]   **17.** $\dfrac{\sqrt[3]{4}}{2}$ [6–5]   **18.** $\dfrac{\sqrt{21}}{7}$ [6–7]   **19.** $\dfrac{5\sqrt{2}+5\sqrt{x}}{2-x}$ [6–4]   **20.** $-3b^2\sqrt[3]{a}$ [6–5]

Chapter 7 Survey

1. $\left\{-4,\dfrac{1}{3}\right\}$ [7–1]   **2.** $\left\{0,\dfrac{5}{3}\right\}$ [7–2]   **3.** $\left\{\pm\sqrt{2}\right\}$ [7–2]   **4.** $\left\{\dfrac{10\pm\sqrt{85}}{5}\right\}$ [7–3]   **5.** $\left\{\dfrac{-4\pm\sqrt{31}}{3}\right\}$ [7–4]   **6.** no real roots [7–6]   **7.** $\dfrac{-1}{5}+\dfrac{13}{5}i$ [7–7]   **8.** $\left\{\pm\sqrt{2},\pm\dfrac{\sqrt{6}}{3}\right\}$ [7–9]   **9.** $\{8\}$ [7–8]   **10.** length: 12 inches; width: 5 inches [7–5]

Exercise 7-1-1

1. $x^2-3x+2=0$ **3.** $6x^2-5x+1=0$ **5.** $2x^2-5x+1=0$ **7.** $5x^2+5x-2=0$

Exercise 7-1-2

1. $\{-1,-2\}$ **3.** $\{-7,-1\}$ **5.** $\{-4,2\}$ **7.** $\{3,6\}$ **9.** $\{-7,3\}$ **11.** $\{-3,1\}$ **13.** $\{-1\}$ **15.** $\{-3,5\}$ **17.** $\left\{-4,\dfrac{3}{2}\right\}$ **19.** $\{-2,3\}$

21. $\{-5, -2\}$ **23.** $\left\{-2, -\dfrac{5}{6}\right\}$

Exercise 7-2-1

1. $\{-3, 0\}$ **3.** $\left\{0, \dfrac{3}{2}\right\}$ **5.** $\{0, 8\}$ **7.** $\left\{0, \dfrac{5}{3}\right\}$ **9.** $\{0, 2\}$ **11.** $\{\pm 2\}$ **13.** $\{\pm\sqrt{5}\}$ **15.** $\{\pm 2\sqrt{5}\}$ **17.** $\{\pm\sqrt{14}\}$ **19.** no real solution

21. $\left\{\pm\dfrac{\sqrt{7}}{2}\right\}$ **23.** $\{-3, 1\}$ **25.** $\left\{\dfrac{-1\pm\sqrt{5}}{3}\right\}$ **27.** $\left\{-\dfrac{2}{5}, 0\right\}$ **29.** $\{\pm\sqrt{7}\}$

Exercise 7-3-1

1. $16; (x+4)^2$ **3.** $121; (x-11)^2$ **5.** $\dfrac{81}{4}; \left(x-\dfrac{9}{2}\right)^2$ **7.** $\dfrac{25}{4}; \left(x+\dfrac{5}{2}\right)^2$

Exercise 7-3-2

1. $\{-5, 1\}$ **3.** $\{1, 7\}$ **5.** $\left\{\dfrac{-3\pm\sqrt{13}}{2}\right\}$ **7.** $\{2, -1\}$ **9.** $\left\{-2, \dfrac{1}{2}\right\}$ **11.** $\left\{\dfrac{-3\pm\sqrt{21}}{3}\right\}$ **13.** no real solution **15.** $\left\{\dfrac{-3\pm 2\sqrt{3}}{2}\right\}$

Exercise 7-4-1

1. $\{-5, 3\}$ **3.** $\left\{-\dfrac{3}{5}, 2\right\}$ **5.** $\left\{\dfrac{-3\pm\sqrt{5}}{2}\right\}$ **7.** $\left\{\dfrac{2\pm\sqrt{2}}{2}\right\}$ **9.** $\left\{\dfrac{1\pm\sqrt{22}}{3}\right\}$ **11.** $\left\{\dfrac{-3\pm\sqrt{3}}{3}\right\}$ **13.** $\left\{\dfrac{3\pm\sqrt{3}}{2}\right\}$ **15.** $\{5\}$ **17.** $\left\{-\dfrac{2}{3}\right\}$

19. no real solution **21.** $\{\pm 2\}$ **23.** $\{5\pm\sqrt{3}\}$ **25.** no real solution **27.** $\{-3, 2\}$ **29.** $\{0, 5\}$

Exercise 7-5-1

1. 3 feet by 8 feet **3.** Width: 6 cm; Length: 8 cm **5.** 5 cm **7.** 8 inches by 12 inches **9.** 12 sides **11.** 1 second and 2 seconds; because the ball passes the specified height both ascending and descending **13.** 450 mph **15.** 4 hours

Exercise 7-6-1

1. -11; no real roots **3.** 25; two rational and unequal roots **5.** 1; two rational and unequal roots **7.** -3; no real roots

Exercise 7-7-1

1. $4i$ **3.** $\sqrt{30}\,i$ **5.** $5\sqrt{2}\,i$ **7.** $10i$

Exercise 7-7-2

1. $7+7i$ **3.** 13 **5.** $8+3i$ **7.** $-10+11i$ **9.** $9+40i$ **11.** $\dfrac{-4}{5}+\dfrac{7}{5}i$ **13.** $\dfrac{2}{37}+\dfrac{12}{37}i$ **15.** $\dfrac{14}{13}-\dfrac{5}{13}i$

Exercise 7-7-3

1. $\{2\pm 2i\}$ **3.** $\left\{\dfrac{1}{4}\pm\dfrac{1}{4}i\right\}$ **5.** $\left\{\dfrac{1}{2}\pm\dfrac{\sqrt{3}}{2}i\right\}$ **7.** $\left\{\dfrac{-1}{6}\pm\dfrac{\sqrt{11}}{6}i\right\}$ **9.** $\left\{\dfrac{1}{5}\pm\dfrac{\sqrt{14}}{5}i\right\}$

Exercise 7-8-1

1. $\{36\}$ **3.** $\left\{\dfrac{23}{3}\right\}$ **5.** $\{9\}$ **7.** $\{0, 4\}$ **9.** $\{0, 3\}$ **11.** no solution

Exercise 7-9-1

1. $\{\pm 1, \pm 2\}$ **3.** $\{\pm\sqrt{2}, \pm 2\}$ **5.** $\left\{\dfrac{2}{3}, 1\right\}$ **7.** $\{4, 16\}$ **9.** $\{25\}$

Chapter 7 Review

1. $\{-6, 5\}$ [7–1]   **2.** $\{3, 8\}$ [7–1]   **3.** $\left\{-\dfrac{3}{2}, \dfrac{1}{3}\right\}$ [7–1]   **4.** $\{\pm\sqrt{17}\}$ [7–2]   **5.** $\{\pm 2\sqrt{5}\}$ [7–2]   **6.** $\left\{0, \dfrac{2}{3}\right\}$ [7–2]   **7.** $\{-5, -3\}$ [7–3]

8. $\{5 \pm \sqrt{17}\}$ [7–3]   **9.** $\left\{\dfrac{-3 \pm \sqrt{13}}{2}\right\}$ [7–3]   **10.** $\{-7, 4\}$ [7–4]   **11.** $\left\{\dfrac{-4 \pm \sqrt{2}}{2}\right\}$ [7–4]   **12.** $\left\{\dfrac{1}{2} \pm \dfrac{\sqrt{3}}{2}i\right\}$ [7–4]

13. 17; two irrational and unequal roots [7–6]   **14.** 81; two rational and unequal roots [7–6]   **15.** −7; two complex roots [7–6]

16. $4 + 3i$ [7–7]   **17.** $3 + 3i$ [7–7]   **18.** $17 - i$ [7–7]   **19.** $3 - i$ [7–7]   **20.** $\dfrac{3}{10} + \dfrac{3}{5}i$ [7–7]   **21.** $\left\{-\dfrac{1}{2}, \dfrac{4}{3}\right\}$ [7–1]   **22.** $\{2 \pm i\}$ [7–7]

23. $\left\{\dfrac{-2 \pm \sqrt{6}}{2}\right\}$ [7–4]   **24.** $\left\{-1, \dfrac{5}{2}\right\}$ [7–1]   **25.** $\{\pm 1, \pm 3\}$ [7–9]   **26.** $\{1\}$ [7–9]   **27.** $\{4, 5\}$ [7–9]   **28.** $\{1, 4\}$ [7–9]   **29.** $\{8\}$ [7–8]

30. $\{3\}$ [7–8]   **31.** width = 7 meters; length = 15 meters [7–5]   **32.** 10 meters and 24 meters [7–5]   **33.** 80 miles per hour [7–5]

34. 6 hours [7–5]

Chapter 7 Practice Test

1. 17; two irrational and unequal roots [7–6]   **2.** $\dfrac{9}{4}$ [7–3]   **3.** $\{\pm\sqrt{10}\}$ [7–2]   **4.** $\{\pm 2\}$ [7–2]   **5.** $\{0, 7\}$ [7–2]   **6.** $\{-3, 1\}$ [7–1]

7. $\{2 \pm \sqrt{3}\}$ [7–4]   **8.** $\left\{-3, \dfrac{1}{2}\right\}$ [7–1]   **9.** $\left\{\dfrac{3}{2} \pm \dfrac{\sqrt{7}}{2}i\right\}$ [7–7]   **10.** $\{\pm\sqrt{5}, \pm\sqrt{2}\,i\}$ [7–9]   **11.** $\{18\}$ [7–8]   **12.** $\{4\}$ [7–8]

13. 7 m [7–5]

Chapter 8 Survey

1.

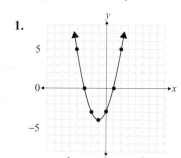

[8–1]   **2. a.** $x \geq -1$ [8–2]   **3.** −1 [8–3]   **4.** $3x + 7y = 2$ [8–3]   **5.** $2x - y = -7$ [8–3]

6. $3x - 2y = -11$ [8–3]   **7.** $(-1, 6)$ and $(2, 3)$

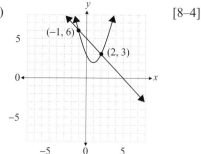

[8–4]

8. $(-3, 4)$ [8–5]   **9.**

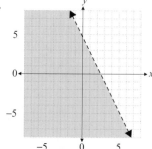

[8–6]   **10.**

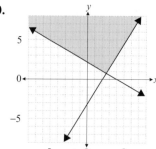

[8–7]   **11.** $(-2, 1, 0)$ [8–8]

Exercise 8-1-1

1. $\{(1, 4), (3, 6), (5, 8)\}$; $y = x + 3$

Exercise 8-1-2

1.

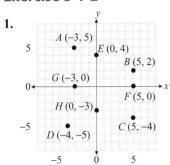

Exercise 8-1-3

1.
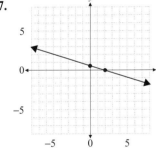

3.
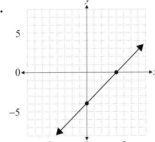

5.
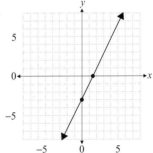

7.
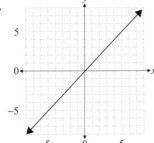

9.
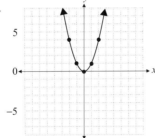

11.
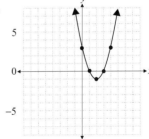

13.
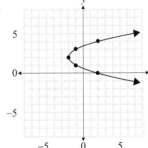

15.
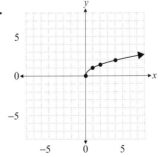

Exercise 8-2-1

1. yes; Domain: $\{-1, 0, 2, 3\}$; Range: $\{-1, 1, 2, 3\}$ **3.** no; Domain: $\{-2, -1, 1\}$; Range: $\{-2, 0, 2, 3\}$ **5.** yes; all real numbers

7. yes; all real numbers   **9.** yes; all real numbers except 5   **11.** no; $x \geq -1$ **13.** yes   **15.** no

Exercise 8-3-1

1. −2

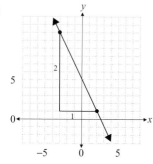

3. $\dfrac{5}{12}$

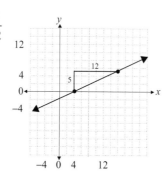

5. 0

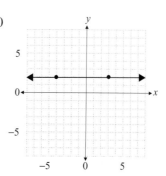

Exercise 8-3-2

1. $3x - y = -4$ **3.** $8x + y = -31$ **5.** $y = 5$ **7.** $2x - 3y = 6$ **9.** $8x - y = 15$ **11.** $5x + 12y = 20$ **13.** $6x - 11y = -37$ **15.** $y = -3$

17. $12x + 6y = 11$

Exercise 8-3-3

1. $m = \dfrac{1}{4}$; y-int: $(0, -5)$

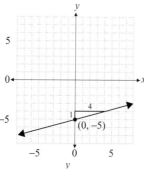

3. $m = \dfrac{2}{5}$; y-int: $(0, -3)$

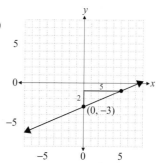

5. $m = -2$; y-int: $(0, 5)$

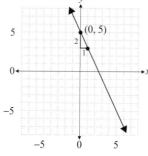

Exercise 8-3-4

1. $2x - y = 5$ **3.** $y = 1$ **5.** $3x + 2y = 13$ **7.** $x = -1$ **9.** $7x - 2y = -14$ **11.** $y = 2$ **13.** $x - 2y = 4$

Exercise 8-4-1

1. $(1, 1)$

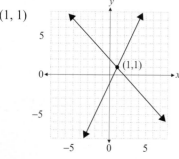

3. $(-2, -3)$

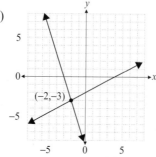

5. $(1, 2)$

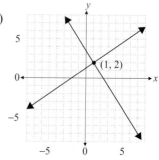

7. (−2, 1) and (0, −1)

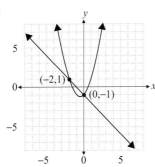

9. No solution - lines are parallel, One solution - one point of intersection, Infinite number of solutions - lines coincide

Exercise 8-5-1

1. (2, 1) **3.** (1, −1) **5.** $\left(\frac{1}{2}, 4\right)$ **7.** $\left(8, \frac{7}{2}\right)$ **9.** (3, −10)

Exercise 8-5-2

1. (3, −1) **3.** (−1, 4) **5.** (2, 1) **7.** (2, −8) **9.** (−2, 5)

Exercise 8-5-3

1. 21 women; 27 men **3.** Brooklyn Bridge:1595 feet, Golden Gate Bridge: 4200 feet **5.** speed of airliner: 805 km/hr, speed of wind: 161 km/hr **7.** The rate of the boat in still water is 9 km/hr. The rate of the current is 3 km/hr.
9. Mike has 16 nickels and 18 dimes.

Exercise 8-5-4

1. independent, (2, −3) **3.** inconsistent **5.** dependent

Exercise 8-6-1

1.
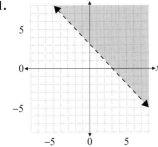

3.
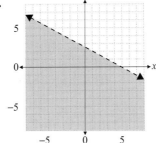

5.
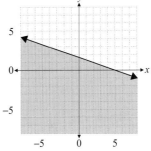

7.
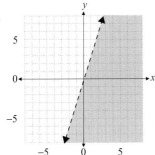

9.
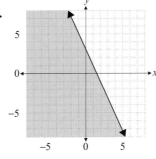

Exercise 8-7-1

1. 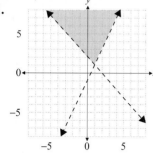 **3.** 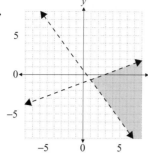 **5.**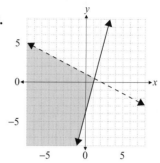

7. 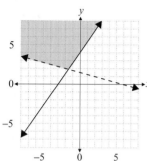 **9.**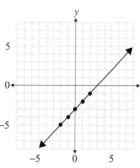

Exercise 8-8-1

1. (1, 2, 3) **3.** (−2, 1, 3) **5.** (2, 1, −3) **7.** (−2, 1, 0) **9.** 5 nickels, 10 dimes, 13 quarters **11.** 2 liters 5%, 3 liters 20%, 4 liters 50%

Chapter 8 Review

1. y: −5, −4, −3, −2, −1 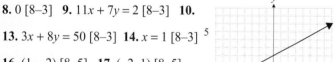 [8–1] **2.** y: 3, 0, −1, 0, 3   [8–1]

3. yes; all real numbers [8–2] **4.** no [8–2] **5.** yes; all real numbers except −1 and 1 [8–2] **6.** $-\dfrac{1}{2}$ [8–3] **7.** undefined [8–3]

8. 0 [8–3] **9.** $11x + 7y = 2$ [8–3] **10.** 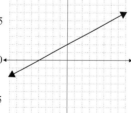 [8–3] **11.** $m = \dfrac{2}{5}$; y-int: (0, −2) [8–3] **12.** $x − 2y = −6$ [8–3]

13. $3x + 8y = 50$ [8–3] **14.** $x = 1$ [8–3] **15.** (0, −1), (−1, 0) 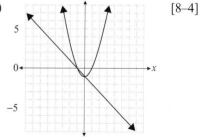 [8–4]

16. (1, −2) [8–5] **17.** (−2, 1) [8–5]

18. (3, 2) [8–5]

19. Speed of boat: 19 km/hr;

Speed of current: 4 km/hr [8–5]

20. Independent; (2, −1) [8–5] **21.** Inconsistent [8–5] **22.** Dependent [8–5]

23. 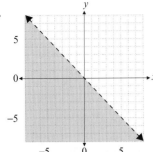 [8–6] **24.** 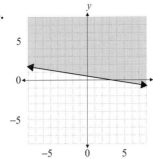 [8–6] **25.** 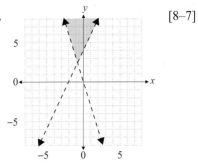 [8–7]

26. 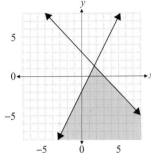 [8–7] **27.** (1, 2, −1) [8–8] **28.** (1, −1, 3) [8–8] **29.** 6 liters [8–8]

Chapter 8 Practice Test

1. yes; $x \geq -2$ [8–2] **2. a.** y: 2, −1, −2, −1, 2 **b.** 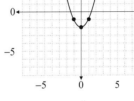 [8–1] **3.** $3x - 5y = 10$ [8–3]

4. $m = \dfrac{2}{3}$; y-int: $\left(0, -\dfrac{1}{3}\right)$ [8–3] **5.** $x - 5y = 10$ [8–3]

6. (−1, 0) and (2, 3) 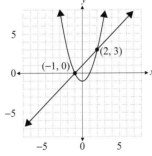 [8–4]

7. (3, 1) 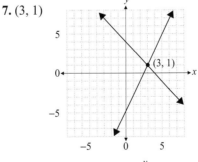 [8–4] **8.** (3, −4) [8–5]

9. (5, −4) [8–5] **10.** 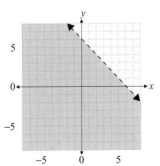 [8–6] **11.** 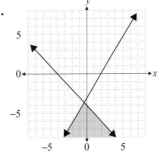 [8–7] **12.** (2, −1, 1) [8–8]

13. first number = 1, second number = −7, third number = 6 [8–8]

End of Book Test

1. -2 [1–3] **2.** $-8a^2$ [1–4] **3.** $4a^5b^4 - 6a^2b^4$ [1–6] **4.** -23 [1–7] **5.** $17x^2 - 4x$ [1–7] **6.** 4 [2–3] **7.** 4 [2–3] **8.** $x = \dfrac{-9a}{4}$

[2–3] **9.** $x \geq 3$ [2–5] **10.** $x < -4 \text{ or } x > 10$ [2–6] **11.** $3ab^2\left(3a - 2b - 4a^2\right)$ [4–2] **12.** $(2x + 5)\left(4x^2 - 10x + 25\right)$ [4–6]

13. $(2a - 3)(x - 10)$ [4–3] **14.** $4(2p - 1)(p + 3)$ [4–7] **15.** $\dfrac{x + 7}{x + 5}$ [5–1] **16.** $\dfrac{-3}{y + 4}$ [5–2] **17.** $\dfrac{9x + 15}{x(x + 3)}$ [5–4]

18. $\dfrac{-2m^2 - 5m + 2}{(m + 5)(m - 1)}$ [5–5] **19.** $\dfrac{1}{x(x + 3)}$ [5–6] **20.** -2 [5–7] **21.** $\dfrac{1}{64}$ [6–4] **22.** -3 [6–3] **23.** $3\sqrt{3}\,x^2$ [6–5] **24.** $3\sqrt{7}$ [6–6]

25. $16 + \sqrt{6}$ [6–6] **26.** $\sqrt{x} + 2$ [6–7] **27.** $-27x^3y^{12}$ [6–1] **28.** $-72a^7b^{11}$ [6–1] **29.** $\dfrac{25x^8}{9y^4}$ [6–2] **30.** $a^{\frac{1}{2}}$ [6–4] **31.** $\dfrac{3}{4xy^2}$

[6–4] **32.** $7 + 17i$ [7–7] **33.** $\left\{\pm\dfrac{2\sqrt{3}}{3}\right\}$ [7–2] **34.** $\left\{-\dfrac{1}{2}, 5\right\}$ [7–1] **35.** $\left\{\dfrac{7}{2}\right\}$ [7–1] **36.** $\{2\}$ [7–8] **37.** $5x - 4y = 35$ [8–3]

38. $(3, -1)$ 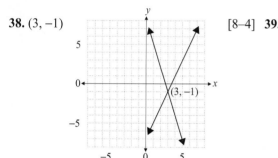 [8–4] **39.** $(-3, 4)$ [8–5] **40.** 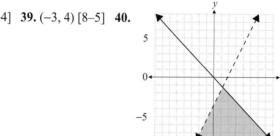 [8–6] **41.** $(2, 1, -2)$ [8–8]

42. Amanda: \$906; Burton: \$883; Carlos: \$1766 [3–2] **43.** 20 mph [3–3] **44.** 4 hours [5–7] **45.** 10 ft by 30 ft [7–5]

INDEX